TREATISE
ON LIGHT

TREATISE ON LIGHT

In which are explained
The causes of that which occurs
In REFLEXION, & in REFRACTION
And particularly
In the strange REFRACTION
OF ICELAND CRYSTAL

By CHRISTIAAN HUYGENS

Rendered into English
BY SILVANUS P. THOMPSON

THE UNIVERSITY OF CHICAGO PRESS
CHICAGO · ILLINOIS

THE UNIVERSITY OF CHICAGO PRESS, CHICAGO 37
Cambridge University Press, London, N.W. 1, England
The University of Toronto Press, Toronto 5, Canada

Published 1945. Fourth Impression 1960

PRINTED IN THE U.S.A.

PREFACE

WROTE this Treatise during my sojourn in France twelve years ago, and I communicated it in the year 1678. to the learned persons who then composed the Royal Academy of Science, to the membership of which the King had done me the honour of calling me. Several of that body who are still alive will remember having been present when I read it, and above the rest those amongst them who applied themselves particularly to the study of Mathematics; of whom I cannot cite more than the celebrated gentlemen Cassini, Römer, and De la Hire. And although I have since corrected and changed some parts, the copies which I had made of it at that time may serve for proof that I have yet added nothing to it save some conjectures touching the formation of Iceland Crystal, and a novel observation on the refraction of Rock Crystal. I have desired to relate these particulars to make known how long I have meditated the things which now I publish, and not for the purpose of detracting from the merit of those who, without having seen anything that I have written, may be found to have treated

of

of like matters: as has in fact occurred to two eminent Geometricians, Messieurs Newton and Leibnitz, with respect to the Problem of the figure of glasses for collecting rays when one of the surfaces is given.

One may ask why I have so long delayed to bring this work to the light. The reason is that I wrote it rather carelessly in the Language in which it appears, with the intention of translating it into Latin, so doing in order to obtain greater attention to the thing. After which I proposed to myself to give it out along with another Treatise on Dioptrics, in which I explain the effects of Telescopes and those things which belong more to that Science. But the pleasure of novelty being past, I have put off from time to time the execution of this design, and I know not when I shall ever come to an end if it, being often turned aside either by business or by some new study. Considering which I have finally judged that it was better worth while to publish this writing, such as it is, than to let it run the risk, by waiting longer, of remaining lost.

There will be seen in it demonstrations of those kinds which do not produce as great a certitude as those of Geometry, and which even differ much therefrom, since whereas the Geometers prove their Propositions by fixed and incontestable Principles, here the Principles are verified by the conclusions to be drawn from them; the nature of these things not allowing of this being done otherwise. It is always possible to attain thereby to a degree of probability which very often is scarcely less than complete proof. To wit, when things which have been demonstrated by the Principles that have been assumed correspond perfectly to the phenomena which experiment has brought under observation; especially when there are a great number of them,

PREFACE

them, and further, principally, when one can imagine and foresee new phenomena which ought to follow from the hypotheses which one employs, and when one finds that therein the fact corresponds to our prevision. But if all these proofs of probability are met with in that which I propose to discuss, as it seems to me they are, this ought to be a very strong confirmation of the success of my inquiry; and it must be ill if the facts are not pretty much as I represent them. I would believe then that those who love to know the Causes of things and who are able to admire the marvels of Light, will find some satisfaction in these various speculations regarding it, and in the new explanation of its famous property which is the main foundation of the construction of our eyes and of those great inventions which extend so vastly the use of them. I hope also that there will be some who by following these beginnings will penetrate much further into this question than I have been able to do, since the subject must be far from being exhausted. This appears from the passages which I have indicated where I leave certain difficulties without having resolved them, and still more from matters which I have not touched at all, such as Luminous Bodies of several sorts, and all that concerns Colours; in which no one until now can boast of having succeeded. Finally, there remains much more to be investigated touching the nature of Light which I do not pretend to have disclosed, and I shall owe much in return to him who shall be able to supplement that which is here lacking to me in knowledge. The Hague. The 8 January 1690.

NOTE BY THE TRANSLATOR

ONSIDERING the great influence which this Treatise has exercised in the development of the Science of Optics, it seems strange that two centuries should have passed before an English edition of the work appeared. Perhaps the circumstance is due to the mistaken zeal with which formerly everything that conflicted with the cherished ideas of Newton was denounced by his followers. The Treatise on Light of Huygens has, however, withstood the test of time: and even now the exquisite skill with which he applied his conception of the propagation of waves of light to unravel the intricacies of the phenomena of the double refraction of crystals, and of the refraction of the atmosphere, will excite the admiration of the student of Optics. It is true that his wave theory was far from the complete doctrine as subsequently developed by Thomas Young and Augustin Fresnel, and belonged rather to geometrical than to physical Optics. If Huygens had no conception of transverse vibrations, of the principle of interference, or of the existence of the ordered sequence of waves in trains, he nevertheless attained to a remarkably clear understanding of the principles

ciples of wave-propagation; and his exposition of the subject marks an epoch in the treatment of Optical problems. It has been needful in preparing this translation to exercise care lest one should import into the author's text ideas of subsequent date, by using words that have come to imply modern conceptions. Hence the adoption of as literal a rendering as possible. A few of the author's terms need explanation. He uses the word "refraction," for example, both for the phenomenon or process usually so denoted, and for the result of that process: thus the refracted ray he habitually terms "the refraction" of the incident ray. When a wave-front, or, as he terms it, a "wave," has passed from some initial position to a subsequent one, he terms the wave-front in its subsequent position "the continuation" of the wave. He also speaks of the envelope of a set of elementary waves, formed by coalescence of those elementary wave-fronts, as "the termination" of the wave; and the elementary wave-fronts he terms "particular" waves. Owing to the circumstance that the French word *rayon* possesses the double signification of ray of light and radius of a circle, he avoids its use in the latter sense and speaks always of the semi-diameter, not of the radius. His speculations as to the ether, his suggestive views of the structure of crystalline bodies, and his explanation of opacity, slight as they are, will possibly surprise the reader by their seeming modernness. And none can read his investigation of the phenomena found in Iceland spar without marvelling at his insight and sagacity.

S. P. T.

June, 1912.

TABLE OF MATTERS
Contained in this Treatise

CHAP. I. On Rays Propagated in Straight Lines.
That Light is produced by a certain movement. *p.* 3
That no substance passes from the luminous object to the eyes. *p.* 3
That Light spreads spherically, almost as Sound does. *p.* 4
Whether Light takes time to spread. *p.* 4
Experience seeming to prove that it passes instantaneously. *p.* 5
Experience proving that it takes time. *p.* 8
How much its speed is greater than that of Sound. *p.* 10
In what the emission of Light differs from that of Sound. *p.* 10
That it is not the same medium which serves for Light and Sound. *p.* 11
How Sound is propagated. *p.* 12
How Light is propagated. *p.* 14
Detailed Remarks on the propagation of Light. *p.* 15
Why Rays are propagated only in straight lines. *p.* 20
How Light coming in different directions can cross itself. *p.* 22

CHAP. II. On Reflexion.
Demonstration of equality of angles of incidence and reflexion. *p.* 23
Why the incident and reflected rays are in the same plane perpendicular to the reflecting surface. *p.* 25
That it is not needful for the reflecting surface to be perfectly flat to attain equality of the angles of incidence and reflexion. *p.* 27

CHAP. III. On Refraction.
That bodies may be transparent without any substance passing through them. *p.* 29
Proof that the ethereal matter passes through transparent bodies. *p.* 30
How this matter passing through can render them transparent. *p.* 31
That the most solid bodies in appearance are of a very loose texture. *p.* 31
That Light spreads more slowly in water and in glass than in air. *p.* 32
Third hypothesis to explain transparency, and the retardation which Light suffers. *p.* 32
On that which makes bodies opaque. *p.* 34
Demonstration why Refraction obeys the known proportion of Sines. *p.* 35
Why the incident and refracted Rays produce one another reciprocally. *p.* 39
Why Reflexion within a triangular glass prism is suddenly augmented when the Light can no longer penetrate. *p.* 40
That bodies which cause greater Refraction also cause stronger Reflexion. *p.* 42
Demonstration of the Theorem of Mr. Fermat. *p.* 43

CHAP. IV. On the Refraction of the Air.
That the emanations of Light in the air are not spherical. *p.* 45
How consequently some objects appear higher than they are. *p.* 47
How the Sun may appear on the Horizon before he has risen. *p.* 49

TABLE OF MATTERS

That the rays of light become curved in the Air of the Atmosphere, and what effects this produces. *p.* 50

CHAP. V. On the Strange Refraction of Iceland Crystal.
That this Crystal grows also in other countries. *p.* 52
Who first wrote about it. *p.* 53
Description of Iceland Crystal; its substance, shape, and properties. *p.* 53
That it has two different Refractions. *p.* 54
That the ray perpendicular to the surface suffers refraction, and that some rays inclined to the surface pass without suffering refraction. *p.* 55
Observation of the refractions in this Crystal. *p.* 56
That there is a Regular and an Irregular Refraction. *p.* 57
The way of measuring the two Refractions of Iceland Crystal. *p.* 57
Remarkable properties of the Irregular Refraction. *p.* 60
Hypothesis to explain the double Refraction. *p.* 61
That Rock Crystal has also a double Refraction. *p.* 62
Hypothesis of emanations of Light, within Iceland Crystal, of spheroidal form, for the Irregular Refraction. *p.* 63
How a perpendicular ray can suffer Refraction. *p.* 64
How the position and form of the spheroidal emanations in this Crystal can be defined. *p.* 65
Explanation of the Irregular Refraction by these spheroidal emanations. *p.* 67
Easy way to find the Irregular Refraction of each incident ray. *p.* 70
Demonstration of the oblique ray which traverses the Crystal without being refracted. *p.* 73
Other irregularities of Refraction explained. *p.* 76
That an object placed beneath the Crystal appears double, in two images of different heights. *p.* 81
Why the apparent heights of one of the images change on changing the position of the eyes above the Crystal. *p.* 85
Of the different sections of this Crystal which produce yet other refractions, and confirm all this Theory. *p.* 88
Particular way of polishing the surfaces after it has been cut. *p.* 91
Surprising phenomenon touching the rays which pass through two separated pieces; the cause of which is not explained. *p.* 92
Probable conjecture on the internal composition of Iceland Crystal, and of what figure its particles are. *p.* 95
Tests to confirm this conjecture. *p.* 97
Calculations which have been supposed in this Chapter. *p.* 99

CHAP. VI. On the Figures of transparent bodies which serve for Refraction and for Reflexion.
General and easy rule to find these Figures. *p.* 106
Invention of the Ovals of Mr. Des Cartes for Dioptrics. *p.* 109
How he was able to find these Lines. *p.* 114
Way of finding the surface of a glass for perfect refraction, when the other surface is given. *p.* 116
Remark on what happens to rays refracted at a spherical surface. *p.* 123
Remark on the curved line which is formed by reflexion in a spherical concave mirror. *p.* 126

TREATISE ON LIGHT

CHAPTER I
On Rays Propagated in Straight Lines

AS happens in all the sciences in which Geometry is applied to matter, the demonstrations concerning Optics are founded on truths drawn from experience. Such are that the rays of light are propagated in straight lines; that the angles of reflexion and of incidence are equal; and that in refraction the ray is bent according to the law of sines, now so well known, and which is no less certain than the preceding laws.

The majority of those who have written touching the various parts of Optics have contented themselves with presuming these truths. But some, more inquiring, have desired to investigate the origin and the causes, considering these to be in themselves wonderful effects of Nature. In which they advanced some ingenious things, but not however such that the most intelligent folk do not wish for better and more satisfactory explanations. Wherefore I here desire to propound what I have meditated on the subject,

ject, so as to contribute as much as I can to the explanation of this department of Natural Science, which, not without reason, is reputed to be one of its most difficult parts. I recognize myself to be much indebted to those who were the first to begin to dissipate the strange obscurity in which these things were enveloped, and to give us hope that they might be explained by intelligible reasoning. But, on the other hand I am astonished also that even here these have often been willing to offer, as assured and demonstrative, reasonings which were far from conclusive. For I do not find that any one has yet given a probable explanation of the first and most notable phenomena of light, namely why it is not propagated except in straight lines, and how visible rays, coming from an infinitude of diverse places, cross one another without hindering one another in any way.

I shall therefore essay in this book, to give, in accordance with the principles accepted in the Philosophy of the present day, some clearer and more probable reasons, firstly of these properties of light propagated rectilinearly; secondly of light which is reflected on meeting other bodies. Then I shall explain the phenomena of those rays which are said to suffer refraction on passing through transparent bodies of different sorts; and in this part I shall also explain the effects of the refraction of the air by the different densities of the Atmosphere.

Thereafter I shall examine the causes of the strange refraction of a certain kind of Crystal which is brought from Iceland. And finally I shall treat of the various shapes of transparent and reflecting bodies by which rays are collected at a point or are turned aside in various ways. From this it will be seen with what facility, following our new Theory, we find not only the Ellipses, Hyperbolas, and
other

other curves which Mr. Des Cartes has ingeniously invented for this purpose; but also those which the surface of a glass lens ought to possess when its other surface is given as spherical or plane, or of any other figure that may be.

It is inconceivable to doubt that light consists in the motion of some sort of matter. For whether one considers its production, one sees that here upon the Earth it is chiefly engendered by fire and flame which contain without doubt bodies that are in rapid motion, since they dissolve and melt many other bodies, even the most solid; or whether one considers its effects, one sees that when light is collected, as by concave mirrors, it has the property of burning as a fire does, that is to say it disunites the particles of bodies. This is assuredly the mark of motion, at least in the true Philosophy, in which one conceives the causes of all natural effects in terms of mechanical motions. This, in my opinion, we must necessarily do, or else renounce all hopes of ever comprehending anything in Physics.

And as, according to this Philosophy, one holds as certain that the sensation of sight is excited only by the impression of some movement of a kind of matter which acts on the nerves at the back of our eyes, there is here yet one reason more for believing that light consists in a movement of the matter which exists between us and the luminous body.

Further, when one considers the extreme speed with which light spreads on every side, and how, when it comes from different regions, even from those directly opposite, the rays traverse one another without hindrance, one may well understand that when we see a luminous object, it cannot be by any transport of matter coming to us from this object,

in

in the way in which a shot or an arrow traverses the air; for assuredly that would too greatly impugn these two properties of light, especially the second of them. It is then in some other way that light spreads; and that which can lead us to comprehend it is the knowledge which we have of the spreading of Sound in the air.

We know that by means of the air, which is an invisible and impalpable body, Sound spreads around the spot where it has been produced, by a movement which is passed on successively from one part of the air to another; and that the spreading of this movement, taking place equally rapidly on all sides, ought to form spherical surfaces ever enlarging and which strike our ears. Now there is no doubt at all that light also comes from the luminous body to our eyes by some movement impressed on the matter which is between the two; since, as we have already seen, it cannot be by the transport of a body which passes from one to the other. If, in addition, light takes time for its passage—which we are now going to examine—it will follow that this movement, impressed on the intervening matter, is successive; and consequently it spreads, as Sound does, by spherical surfaces and waves: for I call them waves from their resemblance to those which are seen to be formed in water when a stone is thrown into it, and which present a successive spreading as circles, though these arise from another cause, and are only in a flat surface.

To see then whether the spreading of light takes time, let us consider first whether there are any facts of experience which can convince us to the contrary. As to those which can be made here on the Earth, by striking lights at great distances, although they prove that light takes no sensible time to pass over these distances, one may say with good reason

ON LIGHT. Chap. I

reason that they are too small, and that the only conclusion to be drawn from them is that the passage of light is extremely rapid. Mr. Des Cartes, who was of opinion that it is instantaneous, founded his views, not without reason, upon a better basis of experience, drawn from the Eclipses of the Moon; which, nevertheless, as I shall show, is not at all convincing. I will set it forth, in a way a little different from his, in order to make the conclusion more comprehensible.

Let A be the place of the sun, BD a part of the orbit or

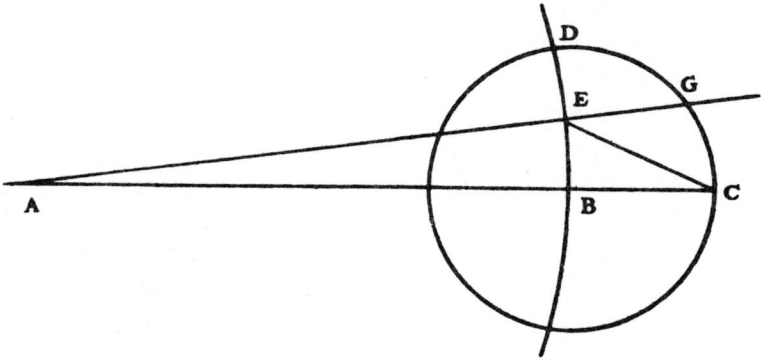

annual path of the Earth: ABC a straight line which I suppose to meet the orbit of the Moon, which is represented by the circle CD, at C.

Now if light requires time, for example one hour, to traverse the space which is between the Earth and the Moon, it will follow that the Earth having arrived at B, the shadow which it casts, or the interruption of the light, will not yet have arrived at the point C, but will only arrive there an hour after. It will then be one hour after, reckoning from the moment when the Earth was at B,

that

that the Moon, arriving at C, will be obscured: but this obscuration or interruption of the light will not reach the Earth till after another hour. Let us suppose that the Earth in these two hours will have arrived at E. The Earth then, being at E, will see the Eclipsed Moon at C, which it left an hour before, and at the same time will see the sun at A. For it being immovable, as I suppose with Copernicus, and the light moving always in straight lines, it must always appear where it is. But one has always observed, we are told, that the eclipsed Moon appears at the point of the Ecliptic opposite to the Sun; and yet here it would appear in arrear of that point by an amount equal to the angle GEC, the supplement of AEC. This, however is contrary to experience, since the angle GEC would be very sensible, and about 33 degrees. Now according to our computation, which is given in the Treatise on the causes of the phenomena of Saturn, the distance BA between the Earth and the Sun is about twelve thousand diameters of the Earth, and hence four hundred times greater than BC the distance of the Moon, which is 30 diameters. Then the angle ECB will be nearly four hundred times greater than BAE, which is five minutes; namely, the path which the earth travels in two hours along its orbit; and thus the angle BCE will be nearly 33 degrees; and likewise the angle CEG, which is greater by five minutes.

But it must be noted that the speed of light in this argument has been assumed such that it takes a time of one hour to make the passage from here to the Moon. If one supposes that for this it requires only one minute of time, then it is manifest that the angle CEG will only be 33 minutes; and if it requires only ten seconds of time, the

the angle will be less than six minutes. And then it will not be easy to perceive anything of it in observations of the Eclipse; nor, consequently, will it be permissible to deduce from it that the movement of light is instantaneous.

It is true that we are here supposing a strange velocity that would be a hundred thousand times greater than that of Sound. For Sound, according to what I have observed, travels about 180 Toises in the time of one Second, or in about one beat of the pulse. But this supposition ought not to seem to be an impossibility; since it is not a question of the transport of a body with so great a speed, but of a successive movement which is passed on from some bodies to others. I have then made no difficulty, in meditating on these things, in supposing that the emanation of light is accomplished with time, seeing that in this way all its phenomena can be explained, and that in following the contrary opinion everything is incomprehensible. For it has always seemed to me that even Mr. Des Cartes, whose aim has been to treat all the subjects of Physics intelligibly, and who assuredly has succeeded in this better than any one before him, has said nothing that is not full of difficulties, or even inconceivable, in dealing with Light and its properties.

But that which I employed only as a hypothesis, has recently received great seemingness as an established truth by the ingenious proof of Mr. Römer which I am going here to relate, expecting him himself to give all that is needed for its confirmation. It is founded as is the preceding argument upon celestial observations, and proves not only that Light takes time for its passage, but also demonstrates how much time it takes, and that its velocity is even at least six times greater than that which I have just stated.
For

8   TREATISE

For this he makes use of the Eclipses suffered by the little planets which revolve around Jupiter, and which often enter his shadow: and see what is his reasoning. Let A be the Sun, BCDE the annual orbit of the Earth, F Jupiter, GN the orbit of the nearest of his Satellites, for it is this one which is more apt for this investigation than any of the other three, because of the quickness of its revolution. Let G be this Satellite entering into the shadow of Jupiter, H the same Satellite emerging from the shadow.

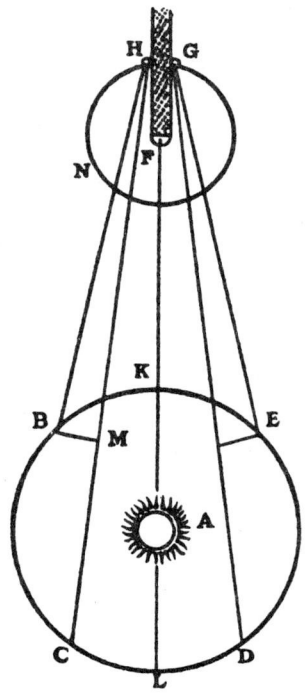

Let it be then supposed, the Earth being at B some time before the last quadrature, that one has seen the said Satellite emerge from the shadow; it must needs be, if the Earth remains at the same place, that, after $42\frac{1}{2}$ hours, one would again see a similar emergence, because that is the time in which it makes the round of its orbit, and when it would come again into opposition to the Sun. And if the Earth, for instance, were to remain always at B during 30 revolutions of this Satellite, one would see it again emerge from the shadow after 30 times $42\frac{1}{2}$ hours. But the Earth having been carried along during this time to C, increasing thus its distance from Jupiter, it follows that if Light requires time for its passage the illumination of the little planet will be perceived later at
                                                                          C

C than it would have been at B, and that there must be added to this time of 30 times $42\tfrac{1}{2}$ hours that which the Light has required to traverse the space MC, the difference of the spaces CH, BH. Similarly at the other quadrature when the earth has come to E from D while approaching toward Jupiter, the immersions of the Satellite ought to be observed at E earlier than they would have been seen if the Earth had remained at D.

Now in quantities of observations of these Eclipses, made during ten consecutive years, these differences have been found to be very considerable, such as ten minutes and more; and from them it has been concluded that in order to traverse the whole diameter of the annual orbit KL, which is double the distance from here to the sun, Light requires about 22 minutes of time.

The movement of Jupiter in his orbit while the Earth passed from B to C, or from D to E, is included in this calculation; and this makes it evident that one cannot attribute the retardation of these illuminations or the anticipation of the eclipses, either to any irregularity occurring in the movement of the little planet or to its eccentricity.

If one considers the vast size of the diameter KL, which according to me is some 24 thousand diameters of the Earth one will acknowledge the extreme velocity of Light. For, supposing that KL is no more than 22 thousand of these diameters, it appears that being traversed in 22 minutes this makes the speed a thousand diameters in one minute, that is $16\tfrac{2}{3}$ diameters in one second or in one beat of the pulse, which makes more than 11 hundred times a hundred thousand toises; since the diameter of the Earth contains 2,865 leagues, reckoned at 25 to the degree, and each

each league is 2,282 Toises, according to the exact measurement which Mr. Picard made by order of the King in 1669. But Sound, as I have said above, only travels 180 toises in the same time of one second: hence the velocity of Light is more than six hundred thousand times greater than that of Sound. This, however, is quite another thing from being instantaneous, since there is all the difference between a finite thing and an infinite. Now the successive movement of Light being confirmed in this way, it follows, as I have said, that it spreads by spherical waves, like the movement of Sound.

But if the one resembles the other in this respect, they differ in many other things; to wit, in the first production of the movement which causes them; in the matter in which the movement spreads; and in the manner in which it is propagated. As to that which occurs in the production of Sound, one knows that it is occasioned by the agitation undergone by an entire body, or by a considerable part of one, which shakes all the contiguous air. But the movement of the Light must originate as from each point of the luminous object, else we should not be able to perceive all the different parts of that object, as will be more evident in that which follows. And I do not believe that this movement can be better explained than by supposing that all those of the luminous bodies which are liquid, such as flames, and apparently the sun and the stars, are composed of particles which float in a much more subtle medium which agitates them with great rapidity, and makes them strike against the particles of the ether which surrounds them, and which are much smaller than they. But I hold also that in luminous solids such as charcoal or metal made red hot in the fire, this same movement is caused by the violent
agitatior

agitation of the particles of the metal or of the wood; those of them which are on the surface striking similarly against the ethereal matter. The agitation, moreover, of the particles which engender the light ought to be much more prompt and more rapid than is that of the bodies which cause sound, since we do not see that the tremors of a body which is giving out a sound are capable of giving rise to Light, even as the movement of the hand in the air is not capable of producing Sound.

Now if one examines what this matter may be in which the movement coming from the luminous body is propagated, which I call Ethereal matter, one will see that it is not the same that serves for the propagation of Sound. For one finds that the latter is really that which we feel and which we breathe, and which being removed from any place still leaves there the other kind of matter that serves to convey Light. This may be proved by shutting up a sounding body in a glass vessel from which the air is withdrawn by the machine which Mr. Boyle has given us, and with which he has performed so many beautiful experiments. But in doing this of which I speak, care must be taken to place the sounding body on cotton or on feathers, in such a way that it cannot communicate its tremors either to the glass vessel which encloses it, or to the machine; a precaution which has hitherto been neglected. For then after having exhausted all the air one hears no Sound from the metal, though it is struck.

One sees here not only that our air, which does not penetrate through glass, is the matter by which Sound spreads; but also that it is not the same air but another kind of matter in which Light spreads; since if the air is removed

removed from the vessel the Light does not cease to traverse it as before.

And this last point is demonstrated even more clearly by the celebrated experiment of Torricelli, in which the tube of glass from which the quicksilver has withdrawn itself, remaining void of air, transmits Light just the same as when air is in it. For this proves that a matter different from air exists in this tube, and that this matter must have penetrated the glass or the quicksilver, either one or the other, though they are both impenetrable to the air. And when, in the same experiment, one makes the vacuum after putting a little water above the quicksilver, one concludes equally that the said matter passes through glass or water, or through both.

As regards the different modes in which I have said the movements of Sound and of Light are communicated, one may sufficiently comprehend how this occurs in the case of Sound if one considers that the air is of such a nature that it can be compressed and reduced to a much smaller space than that which it ordinarily occupies. And in proportion as it is compressed the more does it exert an effort to regain its volume; for this property along with its penetrability, which remains notwithstanding its compression, seems to prove that it is made up of small bodies which float about and which are agitated very rapidly in the ethereal matter composed of much smaller parts. So that the cause of the spreading of Sound is the effort which these little bodies make in collisions with one another, to regain freedom when they are a little more squeezed together in the circuit of these waves than elsewhere.

But the extreme velocity of Light, and other properties which it has, cannot admit of such a propagation of motion, and

and I am about to show here the way in which I conceive it must occur. For this, it is needful to explain the property which hard bodies must possess to transmit movement from one to another.

When one takes a number of spheres of equal size, made of some very hard substance, and arranges them in a straight line, so that they touch one another, one finds, on striking with a similar sphere against the first of these spheres, that the motion passes as in an instant to the last of them, which separates itself from the row, without one's being able to perceive that the others have been stirred. And even that one which was used to strike remains motionless with them. Whence one sees that the movement passes with an extreme velocity which is the greater, the greater the hardness of the substance of the spheres.

But it is still certain that this progression of motion is not instantaneous, but successive, and therefore must take time. For if the movement, or the disposition to movement, if you will have it so, did not pass successively through all these spheres, they would all acquire the movement at the same time, and hence would all advance together; which does not happen. For the last one leaves the whole row and acquires the speed of the one which was pushed. Moreover there are experiments which demonstrate that all the bodies which we reckon of the hardest kind, such as quenched steel, glass, and agate, act as springs and bend somehow, not only when extended as rods but also when they are in the form of spheres or of other shapes. That is to say they yield a little in themselves at the place where they are struck, and immediately regain their former figure. For I have found that on striking with a ball of glass or of agate against a large and quite thick

thick piece of the same substance which had a flat surface, slightly soiled with breath or in some other way, there remained round marks, of smaller or larger size according as the blow had been weak or strong. This makes it evident that these substances yield where they meet, and spring back: and for this time must be required.

Now in applying this kind of movement to that which produces Light there is nothing to hinder us from estimating the particles of the ether to be of a substance as nearly approaching to perfect hardness and possessing a springiness as prompt as we choose. It is not necessary to examine here the causes of this hardness, or of that springiness, the consideration of which would lead us too far from our subject. I will say, however, in passing that we may conceive that the particles of the ether, notwithstanding their smallness, are in turn composed of other parts and that their springiness consists in the very rapid movement of a subtle matter which penetrates them from every side and constrains their structure to assume such a disposition as to give to this fluid matter the most overt and easy passage possible. This accords with the explanation which Mr. Des Cartes gives for the spring, though I do not, like him, suppose the pores to be in the form of round hollow canals. And it must not be thought that in this there is anything absurd or impossible, it being on the contrary quite credible that it is this infinite series of different sizes of corpuscles, having different degrees of velocity, of which Nature makes use to produce so many marvellous effects.

But though we shall ignore the true cause of springiness we still see that there are many bodies which possess this property; and thus there is nothing strange in supposing that

ON LIGHT. Chap. I

that it exists also in little invisible bodies like the particles of the Ether. Also if one wishes to seek for any other way in which the movement of Light is successively communicated, one will find none which agrees better, with uniform progression, as seems to be necessary, than the property of springiness; because if this movement should grow slower in proportion as it is shared over a greater quantity of matter, in moving away from the source of the light, it could not conserve this great velocity over great distances. But by supposing springiness in the ethereal matter, its particles will have the property of equally rapid restitution whether they are pushed strongly or feebly; and thus the propagation of Light will always go on with an equal velocity.

And it must be known that although the particles of the ether are not ranged thus in straight lines, as in our row of spheres, but confusedly, so that one of them touches several others, this does not hinder them from transmitting their movement and from spreading it always forward. As to this it is to be remarked that there is a law of motion serving for this propagation, and verifiable by experiment. It is that when a sphere, such as A here, touches several other similar spheres CCC, if it is struck by another sphere B in such a way as to exert an impulse against all the spheres CCC which touch it, it transmits to them the whole of its movement, and remains 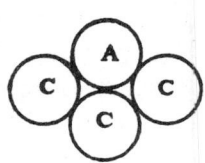 after that motionless like the sphere B. And without supposing that the ethereal particles are of spherical form (for I see indeed no need to suppose them so) one may well understand that this property of communicating an impulse

pulse does not fail to contribute to the aforesaid propagation of movement.

Equality of size seems to be more necessary, because otherwise there ought to be some reflexion of movement backwards when it passes from a smaller particle to a larger one, according to the Laws of Percussion which I published some years ago.

However, one will see hereafter that we have to suppose such an equality not so much as a necessity for the propagation of light as for rendering that propagation easier and more powerful; for it is not beyond the limits of probability that the particles of the ether have been made equal for a purpose so important as that of light, at least in that vast space which is beyond the region of atmosphere and which seems to serve only to transmit the light of the Sun and the Stars.

I have then shown in what manner one may conceive Light to spread successively, by spherical waves, and how it is possible that this spreading is accomplished with as great a velocity as that which experiments and celestial observations demand. Whence it may be further remarked that although the particles are supposed to be in continual movement (for there are many reasons for this) the successive propagation of the waves cannot be hindered by this; because the propagation consists nowise in the transport of those particles but merely in a small agitation which they cannot help communicating to those surrounding, notwithstanding any movement which may act on them causing them to be changing positions amongst themselves.

But we must consider still more particularly the origin of these waves, and the manner in which they spread. And, first, it follows from what has been said on the production

ON LIGHT. Chap. I   17

duction of Light, that each little region of a luminous body, such as the Sun, a candle, or a burning coal, generates its own waves of which that region is the centre. Thus in the flame of a candle, having distinguished the points A, B, C, concentric circles described about each of these points represent the waves which come from them. And one must imagine the same about every point of the surface and of the part within the flame.

But as the percussions at the centres of these waves possess no regular succession, it must not be supposed that the waves themselves follow one another at equal distances: and if the distances marked in the figure appear to be such, it is rather to mark the progression of one and the same wave at equal intervals of time than to represent several of them issuing from one and the same centre.

After all, this prodigious quantity of waves which traverse one another without confusion and without effacing one another must not be deemed inconceivable; it being certain that one and the same particle of matter can serve for many waves coming from different sides or even from contrary directions, not only if it is struck by blows which follow one another closely but even for those which act on it at the same instant. It can do so because the spreading of the movement is successive. This may be proved by the row of equal spheres of hard matter, spoken of above. If against this row there are pushed from two opposite sides at the same time two similar spheres A and

D   D,

D, one will see each of them rebound with the same velocity which it had in striking, yet the whole row will remain in

its place, although the movement has passed along its whole length twice over. And if these contrary movements happen to meet one another at the middle sphere, B, or at some other such as C, that sphere will yield and act as a spring at both sides, and so will serve at the same instant to transmit these two movements.

But what may at first appear full strange and even incredible is that the undulations produced by such small movements and corpuscles, should spread to such immense distances; as for example from the Sun or from the Stars to us. For the force of these waves must grow feeble in proportion as they move away from their origin, so that the action of each one in particular will without doubt become incapable of making itself felt to our sight. But one will cease to be astonished by considering how at a great distance from the luminous body an infinitude of waves, though they have issued from different points of this body, unite together in such a way that they sensibly compose one single wave only, which, consequently, ought to have enough force to make itself felt. Thus this infinite number of waves which originate at the same instant from all points of a fixed star, big it may be as the Sun, make practically only one single wave which may well have force enough to produce an impression on our eyes. Moreover from each luminous point there may come many thousands of waves in the smallest imaginable time, by the frequent percussion of the corpuscles which strike the
Ether

ON LIGHT. Chap. I

Ether at these points: which further contributes to rendering their action more sensible.

There is the further consideration in the emanation of these waves, that each particle of matter in which a wave spreads, ought not to communicate its motion only to the next particle which is in the straight line drawn from the luminous point, but that it also imparts some of it necessarily to all the others which touch it and which oppose themselves to its movement. So it arises that around each particle there is made a wave of which that particle is the centre. Thus if DCF is a wave emanating from the luminous point A, which is its centre, the particle B, one of those comprised within the sphere DCF, will have made its particular or partial wave KCL, which will touch the wave DCF at C at the same moment that the principal wave emanating from the point A has arrived at DCF; and it is clear that it will be only the region C of the wave KCL which will touch the wave DCF, to wit, that which is in the straight line drawn through AB. Similarly the other particles of the sphere DCF, such as *bb*, *dd*, etc., will each make its own wave. But each of these waves can be infinitely feeble only as compared with the wave DCF, to the composition of which all the others contribute by the part of their surface which is most distant from the centre A.

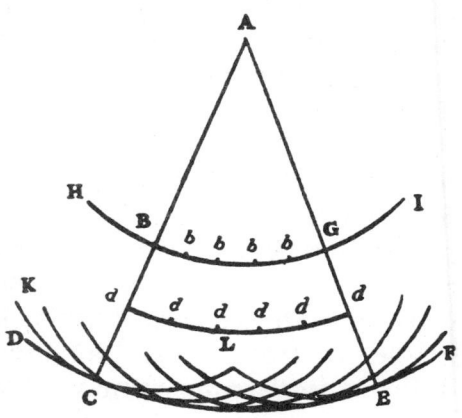

One

One sees, in addition, that the wave DCF is determined by the distance attained in a certain space of time by the movement which started from the point A; there being no movement beyond this wave, though there will be in the space which it encloses, namely in parts of the particular waves, those parts which do not touch the sphere DCF. And all this ought not to seem fraught with too much minuteness or subtlety, since we shall see in the sequel that all the properties of Light, and everything pertaining to its reflexion and its refraction, can be explained in principle by this means. This is a matter which has been quite unknown to those who hitherto have begun to consider the waves of light, amongst whom are Mr. Hooke in his *Micrographia*, and Father Pardies, who, in a treatise of which he let me see a portion, and which he was unable to complete as he died shortly afterward, had undertaken to prove by these waves the effects of reflexion and refraction. But the chief foundation, which consists in the remark I have just made, was lacking in his demonstrations; and for the rest he had opinions very different from mine, as may be will appear some day if his writing has been preserved.

To come to the properties of Light. We remark first that each portion of a wave ought to spread in such a way that its extremities lie always between the same straight lines drawn from the luminous point. Thus the portion BG of the wave, having the luminous point A as its centre, will spread into the arc CE bounded by the straight lines ABC, AGE. For although the particular waves produced by the particles comprised within the space CAE spread also outside this space, they yet do not concur at the same instant to compose a wave which terminates the movement

movement, as they do precisely at the circumference CE, which is their common tangent.

And hence one sees the reason why light, at least if its rays are not reflected or broken, spreads only by straight lines, so that it illuminates no object except when the path from its source to that object is open along such lines. For if, for example, there were an opening BG, limited by opaque bodies BH, GI, the wave of light which issues from the point A will always be terminated by the straight lines AC, AE, as has just been shown; the parts of the partial waves which spread outside the space ACE being too feeble to produce light there.

Now, however small we make the opening BG, there is always the same reason causing the light there to pass between straight lines; since this opening is always large enough to contain a great number of particles of the ethereal matter, which are of an inconceivable smallness; so that it appears that each little portion of the wave necessarily advances following the straight line which comes from the luminous point. Thus then we may take the rays of light as if they were straight lines.

It appears, moreover, by what has been remarked touching the feebleness of the particular waves, that it is not needful that all the particles of the Ether should be equal amongst themselves, though equality is more apt for the propagation of the movement. For it is true that inequality will cause a particle by pushing against another larger one to strive to recoil with a part of its movement; but it will thereby merely generate backwards towards the luminous point some partial waves incapable of causing light, and not a wave compounded of many as CE was.

Another property of waves of light, and one of the most marvellous,

marvellous, is that when some of them come from different or even from opposing sides, they produce their effect across one another without any hindrance. Whence also it comes about that a number of spectators may view different objects at the same time through the same opening, and that two persons can at the same time see one another's eyes. Now according to the explanation which has been given of the action of light, how the waves do not destroy nor interrupt one another when they cross one another, these effects which I have just mentioned are easily conceived. But in my judgement they are not at all easy to explain according to the views of Mr. Des Cartes, who makes Light to consist in a continuous pressure merely tending to movement. For this pressure not being able to act from two opposite sides at the sàme time, against bodies which have no inclination to approach one another, it is impossible so to understand what I have been saying about two persons mutually seeing one another's eyes, or how two torches can illuminate one another.

CHAPTER II
On Reflexion

AVING explained the effects of waves of light which spread in a homogenecus matter, we will examine next that which happens to them on encountering other bodies. We will first make evident how the Reflexion of light is explained by these same waves, and why it preserves equality of angles.

Let

ON·LIGHT. Chap. II

Let there be a surface AB, plane and polished, of some metal, glass, or other body, which at first I will consider as perfectly uniform (reserving to myself to deal at the end of this demonstration with the inequalities from which it cannot be exempt), and let a line AC, inclined to AD, represent a portion of a wave of light, the centre of which is so distant that this portion AC may be considered as a straight line; for I consider all this as in one plane, imagining to myself that the plane in which this figure is, cuts the sphere of the wave through its centre and intersects the plane AB at right angles. This explanation will suffice once for all.

The piece C of the wave AC, will in a certain space of time advance as far as the plane AB at B, following the straight line CB, which may be supposed to come from the luminous centre, and which in consequence is perpendicular to AC. Now in this same space of time the portion A of the same wave, which has been hindered from communicating its movement beyond the plane AB, or at least partly so, ought to have continued its movement in the matter which is above this plane, and this along a distance equal to CB, making its

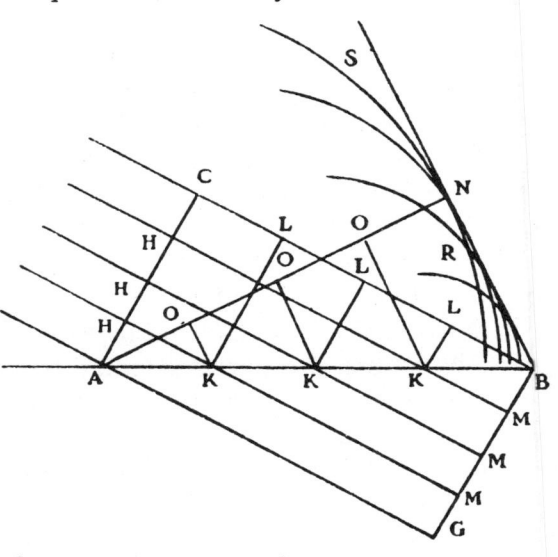

own

own partial spherical wave, according to what has been said above. Which wave is here represented by the circumference SNR, the centre of which is A, and its semi-diameter AN equal to CB.

If one considers further the other pieces H of the wave AC, it appears that they will not only have reached the surface AB by straight lines HK parallel to CB, but that in addition they will have generated in the transparent air, from the centres K, K, K, particular spherical waves, represented here by circumferences the semi-diameters of which are equal to KM, that is to say to the continuations of HK as far as the line BG parallel to AC. But all these circumferences have as a common tangent the straight line BN, namely the same which is drawn from B as a tangent to the first of the circles, of which A is the centre, and AN the semi-diameter equal to BC, as is easy to see.

It is then the line BN (comprised between B and the point N where the perpendicular from the point A falls) which is as it were formed by all these circumferences, and which terminates the movement which is made by the reflexion of the wave AC; and it is also the place where the movement occurs in much greater quantity than anywhere else. Wherefore, according to that which has been explained, BN is the propagation of the wave AC at the moment when the piece C of it has arrived at B. For there is no other line which like BN is a common tangent to all the aforesaid circles, except BG below the plane AB; which line BG would be the propagation of the wave if the movement could have spread in a medium homogeneous with that which is above the plane. And if one wishes to see how the wave AC has come successively to BN, one has only to draw in the same figure the straight lines KO
parallel

parallel to BN, and the straight lines KL parallel to AC. Thus one will see that the straight wave AC has become broken up into all the OKL parts successively, and that it has become straight again at NB.

Now it is apparent here that the angle of reflexion is made equal to the angle of incidence. For the triangles ACB, BNA being rectangular and having the side AB common, and the side CB equal to NA, it follows that the angles opposite to these sides will be equal, and therefore also the angles CBA, NAB. But as CB, perpendicular to CA, marks the direction of the incident ray, so AN, perpendicular to the wave BN, marks the direction of the reflected ray; hence these rays are equally inclined to the plane AB.

But in considering the preceding demonstration, one might aver that it is indeed true that BN is the common tangent of the circular waves in the plane of this figure, but that these waves, being in truth spherical, have still an infinitude of similar tangents, namely all the straight lines which are drawn from the point B in the surface generated by the straight line BN about the axis BA. It remains, therefore, to demonstrate that there is no difficulty herein: and by the same argument one will see why the incident ray and the reflected ray are always in one and the same plane perpendicular to the reflecting plane. I say then that the wave AC, being regarded only as a line, produces no light. For a visible ray of light, however narrow it may be, has always some width, and consequently it is necessary, in representing the wave whose progression constitutes the ray, to put instead of a line AC some plane figure such as the circle HC in the following figure, by supposing, as we have done, the luminous point to be infinitely distant.

Now

Now it is easy to see, following the preceding demonstration, that each small piece of this wave HC having arrived at the plane AB, and there generating each one its particular wave, these will all have, when C arrives at B, a common plane which will touch them, namely a circle BN similar to CH; and this will be intersected at its middle and at right angles by the same plane which likewise intersects the circle CH and the ellipse AB.

One sees also that the said spheres of the partial waves cannot have any common tangent plane other than the

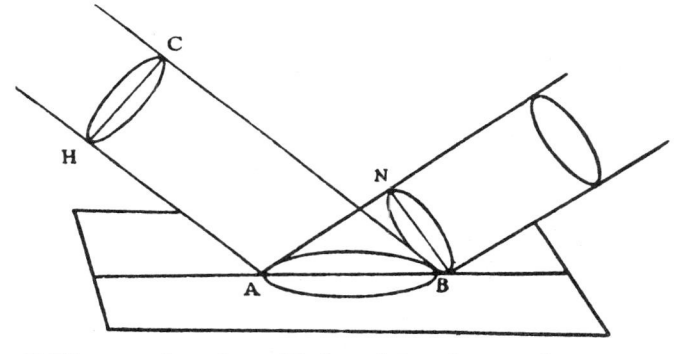

circle BN; so that it will be this plane where there will be more reflected movement than anywhere else, and which will therefore carry on the light in continuance from the wave CH.

I have also stated in the preceding demonstration that the movement of the piece A of the incident wave is not able to communicate itself beyond the plane AB, or at least not wholly. Whence it is to be remarked that though the movement of the ethereal matter might communicate itself partly to that of the reflecting body, this could in nothing alter the velocity of progression of the waves, on which
the

the angle of reflexion depends. For a slight percussion ought to generate waves as rapid as strong percussion in the same matter. This comes about from the property of bodies which act as springs, of which we have spoken above; namely that whether compressed little or much they recoil in equal times. Equally so in every reflexion of the light, against whatever body it may be, the angles of reflexion and incidence ought to be equal notwithstanding that the body might be of such a nature that it takes away a portion of the movement made by the incident light. And experiment shows that in fact there is no polished body the reflexion of which does not follow this rule.

But the thing to be above all remarked in our demonstration is that it does not require that the reflecting surface should be considered as a uniform plane, as has been supposed by all those who have tried to explain the effects of reflexion; but only an evenness such as may be attained by the particles of the matter of the reflecting body being set near to one another; which particles are larger than those of the ethereal matter, as will appear by what we shall say in treating of the transparency and opacity of bodies. For the surface consisting thus of particles put together, and the ethereal particles being above, and smaller, it is evident that one could not demonstrate the equality of the angles of incidence and reflexion by similitude to that which happens to a ball thrown against a wall, of which writers have always made use. In our way, on the other hand, the thing is explained without difficulty. For the smallness of the particles of quicksilver, for example, being such that one must conceive millions of them, in the smallest visible surface proposed, arranged like a heap of grains of sand which has been flattened as much as it is capable of being,
this

this surface then becomes for our purpose as even as a polished glass is: and, although it always remains rough with respect to the particles of the Ether it is evident that the centres of all the particular spheres of reflexion, of which we have spoken, are almost in one uniform plane, and that thus the common tangent can fit to them as perfectly as is requisite for the production of light. And this alone is requisite, in our method of demonstration, to cause equality of the said angles without the remainder of the movement reflected from all parts being able to produce any contrary effect.

CHAPTER III
On Refraction

IN the same way as the effects of Reflexion have been explained by waves of light reflected at the surface of polished bodies, we will explain transparency and the phenomena of refraction by waves which spread within and across diaphanous bodies, both solids, such as glass, and liquids, such as water, oils, etc. But in order that it may not seem strange to suppose this passage of waves in the interior of these bodies, I will first show that one may conceive it possible in more than one mode.

First, then, if the ethereal matter cannot penetrate transparent bodies at all, their own particles would be able to communicate successively the movement of the waves, the same as do those of the Ether, supposing that, like those, they are of a nature to act as a spring. And this is easy

easy to conceive as regards water and other transparent liquids, they being composed of detached particles. But it may seem more difficult as regards glass and other transparent and hard bodies, because their solidity does not seem to permit them to receive movement except in their whole mass at the same time. This, however, is not necessary because this solidity is not such as it appears to us, it being probable rather that these bodies are composed of particles merely placed close to one another and held together by some pressure from without of some other matter, and by the irregularity of their shapes. For primarily their rarity is shown by the facility with which there passes through them the matter of the vortices of the magnet, and that which causes gravity. Further, one cannot say that these bodies are of a texture similar to that of a sponge or of light bread, because the heat of the fire makes them flow and thereby changes the situation of the particles amongst themselves. It remains then that they are, as has been said, assemblages of particles which touch one another without constituting a continuous solid. This being so, the movement which these particles receive to carry on the waves of light, being merely communicated from some of them to others, without their going for that purpose out of their places or without derangement, it may very well produce its effect without prejudicing in any way the apparent solidity of the compound.

By pressure from without, of which I have spoken, must not be understood that of the air, which would not be sufficient, but that of some other more subtle matter, a pressure which I chanced upon by experiment long ago, namely in the case of water freed from air, which remains suspended in a tube open at its lower end, notwithstanding that

that the air has been removed from the vessel in which this tube is enclosed.

One can then in this way conceive of transparency in a solid without any necessity that the ethereal matter which serves for light should pass through it, or that it should find pores in which to insinuate itself. But the truth is that this matter not only passes through solids, but does so even with great facility; of which the experiment of Torricelli, above cited, is already a proof. Because on the quicksilver and the water quitting the upper part of the glass tube, it appears that it is immediately filled with ethereal matter, since light passes across it. But here is another argument which proves this ready penetrability, not only in transparent bodies but also in all others.

When light passes across a hollow sphere of glass, closed on all sides, it is certain that it is full of ethereal matter, as much as the spaces outside the sphere. And this ethereal matter, as has been shown above, consists of particles which just touch one another. If then it were enclosed in the sphere in such a way that it could not get out through the pores of the glass, it would be obliged to follow the movement of the sphere when one changes its place: and it would require consequently almost the same force to impress a certain velocity on this sphere, when placed on a horizontal plane, as if it were full of water or perhaps of quicksilver: because every body resists the velocity of the motion which one would give to it, in proportion to the quantity of matter which it contains, and which is obliged to follow this motion. But on the contrary one finds that the sphere resists the impress of movement only in proportion to the quantity of matter of the glass of which it is made. Then it must be that the ethereal matter which

is

is inside is not shut up, but flows through it with very great freedom. We shall demonstrate hereafter that by this process the same penetrability may be inferred also as relating to opaque bodies.

The second mode then of explaining transparency, and one which appears more probably true, is by saying that the waves of light are carried on in the ethereal matter, which continuously occupies the interstices or pores of transparent bodies. For since it passes through them continuously and freely, it follows that they are always full of it. And one may even show that these interstices occupy much more space than the coherent particles which constitute the bodies. For if what we have just said is true: that force is required to impress a certain horizontal velocity on bodies in proportion as they contain coherent matter; and if the proportion of this force follows the law of weights, as is confirmed by experiment, then the quantity of the constituent matter of bodies also follows the proportion of their weights. Now we see that water weighs only one fourteenth part as much as an equal portion of quicksilver: therefore the matter of the water does not occupy the fourteenth part of the space which its mass obtains. It must even occupy much less of it, since quicksilver is less heavy than gold, and the matter of gold is by no means dense, as follows from the fact that the matter of the vortices of the magnet and of that which is the cause of gravity pass very freely through it.

But it may be objected here that if water is a body of so great rarity, and if its particles occupy so small a portion of the space of its apparent bulk, it is very strange how it yet resists Compression so strongly without permitting itself to be condensed by any force which one has hitherto

hitherto essayed to employ, preserving even its entire liquidity while subjected to this pressure.

This is no small difficulty. It may, however, be resolved by saying that the very violent and rapid motion of the subtle matter which renders water liquid, by agitating the particles of which it is composed, maintains this liquidity in spite of the pressure which hitherto any one has been minded to apply to it.

The rarity of transparent bodies being then such as we have said, one easily conceives that the waves might be carried on in the ethereal matter which fills the interstices of the particles. And, moreover, one may believe that the progression of these waves ought to be a little slower in the interior of bodies, by reason of the small detours which the same particles cause. In which different velocity of light I shall show the cause of refraction to consist.

Before doing so, I will indicate the third and last mode in which transparency may be conceived; which is by supposing that the movement of the waves of light is transmitted indifferently both in the particles of the ethereal matter which occupy the interstices of bodies, and in the particles which compose them, so that the movement passes from one to the other. And it will be seen hereafter that this hypothesis serves excellently to explain the double refraction of certain transparent bodies.

Should it be objected that if the particles of the ether are smaller than those of transparent bodies (since they pass through their intervals), it would follow that they can communicate to them but little of their movement, it may be replied that the particles of these bodies are in turn composed of still smaller particles, and so it will be these

these secondary particles which will receive the movement from those of the ether.

Furthermore, if the particles of transparent bodies have a recoil a little less prompt than that of the ethereal particles, which nothing hinders us from supposing, it will again follow that the progression of the waves of light will be slower in the interior of such bodies than it is outside in the ethereal matter.

All this I have found as most probable for the mode in which the waves of light pass across transparent bodies. To which it must further be added in what respect these bodies differ from those which are opaque; and the more so since it might seem because of the easy penetration of bodies by the ethereal matter, of which mention has been made, that there would not be any body that was not transparent. For by the same reasoning about the hollow sphere which I have employed to prove the smallness of the density of glass and its easy penetrability by the ethereal matter, one might also prove that the same penetrability obtains for metals and for every other sort of body. For this sphere being for example of silver, it is certain that it contains some of the ethereal matter which serves for light, since this was there as well as in the air when the opening of the sphere was closed. Yet, being closed and placed upon a horizontal plane, it resists the movement which one wishes to give to it, merely according to the quantity of silver of which it is made; so that one must conclude, as above, that the ethereal matter which is enclosed does not follow the movement of the sphere; and that therefore silver, as well as glass, is very easily penetrated by this matter. Some of it is therefore present continuously and in quantities between the particles of silver and of all other opaque bodies:

bodies: and since it serves for the propagation of light it would seem that these bodies ought also to be transparent, which however is not the case.

Whence then, one will say, does their opacity come? Is it because the particles which compose them are soft; that is to say, these particles being composed of others that are smaller, are they capable of changing their figure on receiving the pressure of the ethereal particles, the motion of which they thereby damp, and so hinder the continuance of the waves of light? That cannot be: for if the particles of the metals are soft, how is it that polished silver and mercury reflect light so strongly? What I find to be most probable herein, is to say that metallic bodies, which are almost the only really opaque ones, have mixed amongst their hard particles some soft ones; so that some serve to cause reflexion and the others to hinder transparency; while, on the other hand, transparent bodies contain only hard particles which have the faculty of recoil, and serve together with those of the ethereal matter for the propagation of the waves of light, as has been said.

Let us pass now to the explanation of the effects of Refraction, assuming, as we have done, the passage of waves of light through transparent bodies, and the diminution of velocity which these same waves suffer in them.

The chief property of Refraction is that a ray of light, such as AB, being in the air, and falling obliquely upon the polished surface of a transparent body, such as FG, is broken

broken at the point of incidence B, in such a way that
with the straight line DBE which cuts the surface perpendicularly it makes an angle CBE less than ABD which
it made with the same perpendicular when in the air. And
the measure of these angles is found by describing, about
the point B, a circle which cuts the radii AB, BC. For
the perpendiculars AD, CE, let fall from the points of
intersection upon the straight line DE, which are called
the Sines of the angles ABD, CBE, have a certain ratio
between themselves; which ratio is always the same for all
inclinations of the incident ray, at least for a given transparent body. This ratio is, in glass, very nearly as 3 to 2;
and in water very nearly as 4 to 3; and is likewise different
in other diaphanous bodies.

Another property, similar to this, is that the refractions
are reciprocal between the rays entering into a transparent
body and those which are leaving it. That is to say that
if the ray AB in entering the transparent body is refracted into BC, then likewise CB being taken as a ray in the interior of this body will be refracted, on passing out, into BA.

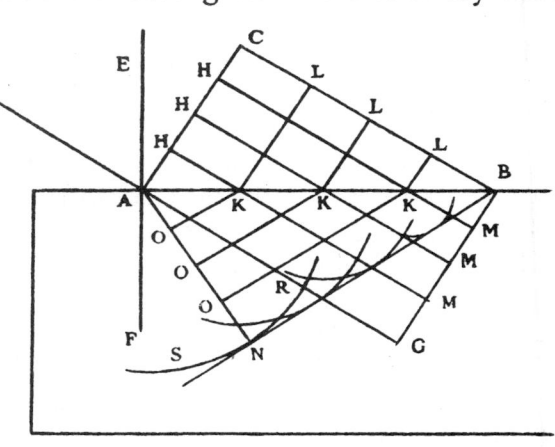

To explain then the reasons of these phenomena according to our principles, let AB be the straight line which

represents

represents a plane surface bounding the transparent substances which lie towards C and towards N. When I say plane, that does not signify a perfect evenness, but such as has been understood in treating of reflexion, and for the same reason. Let the line AC represent a portion of a wave of light, the centre of which is supposed so distant that this portion may be considered as a straight line. The piece C, then, of the wave AC, will in a certain space of time have advanced as far as the plane AB following the straight line CB, which may be imagined as coming from the luminous centre, and which consequently will cut AC at right angles. Now in the same time the piece A would have come to G along the straight line AG, equal and parallel to CB; and all the portion of wave AC would be at GB if the matter of the transparent body transmitted the movement of the wave as quickly as the matter of the Ether. But let us suppose that it transmits this movement less quickly, by one-third, for instance. Movement will then be spread from the point A, in the matter of the transparent body through a distance equal to two-thirds of CB, making its own particular spherical wave according to what has been said before. This wave is then represented by the circumference SNR, the centre of which is A, and its semi-diameter equal to two-thirds of CB. Then if one considers in order the other pieces H of the wave AC, it appears that in the same time that the piece C reaches B they will not only have arrived at the surface AB along the straight lines HK parallel to CB, but that, in addition, they will have generated in the diaphanous substance from the centres K, partial waves, represented here by circumferences the semi-diameters of which are equal to two-thirds of the lines KM, that is to say, to two-thirds two-thirds of the prolongations of HK down to the straight line BG; for these semi-diameters would have been equal to entire lengths of KM if the two transparent substances had been of the same penetrability.

Now all these circumferences have for a common tangent the straight line BN; namely the same line which is drawn as a tangent from the point B to the circumference SNR which we considered first. For it is easy to see that all the other circumferences will touch the same BN, from B up to the point of contact N, which is the same point where AN falls perpendicularly on BN.

It is then BN, which is formed by small arcs of these circumferences, which terminates the movement that the wave AC has communicated within the transparent body, and where this movement occurs in much greater amount than anywhere else. And for that reason this line, in accordance with what has been said more than once, is the propagation of the wave AC at the moment when its piece C has reached B. For there is no other line below the plane AB which is, like BN, a common tangent to all these partial waves. And if one would know how the wave AC has come progressively to BN, it is necessary only to draw in the same figure the straight lines KO parallel to BN, and all the lines KL parallel to AC. Thus one will see that the wave CA, from being a straight line, has become broken in all the positions LKO successively, and that it has again become a straight line at BN. This being evident by what has already been demonstrated, there is no need to explain it further.

Now, in the same figure, if one draws EAF, which cuts the plane AB at right angles at the point A, since AD is perpendicular to the wave AC, it will be DA which will

mark

mark the ray of incident light, and AN which was perpendicular to BN, the refracted ray: since the rays are nothing else than the straight lines along which the portions of the waves advance.

Whence it is easy to recognize this chief property of refraction, namely that the Sine of the angle DAE has always the same ratio to the Sine of the angle NAF, whatever be the inclination of the ray DA: and that this ratio is the same as that of the velocity of the waves in the transparent substance which is towards AE to their velocity in the transparent substance towards AF. For, considering AB as the radius of a circle, the Sine of the angle BAC is BC, and the Sine of the angle ABN is AN. But the angle BAC is equal to DAE, since each of them added to CAE makes a right angle. And the angle ABN is equal to NAF, since each of them with BAN makes a right angle. Then also the Sine of the angle DAE is to the Sine of NAF as BC is to AN. But the ratio of BC to AN was the same as that of the velocities of light in the substance which is towards AE and in that which is towards AF; therefore also the Sine of the angle DAE will be to the Sine of the angle NAF the same as the said velocities of light.

To see, consequently, what the refraction will be when the waves of light pass into a substance in which the movement travels more quickly than in that from which they emerge (let us again assume the ratio of 3 to 2), it is only necessary to repeat all the same construction and demonstration which we have just used, merely substituting everywhere $\frac{3}{2}$ instead of $\frac{2}{3}$. And it will be found by the same reasoning, in this other figure, that when the piece C of the wave AC shall have reached the surface AB at B,
all

all the portions of the wave AC will have advanced as far as BN, so that BC the perpendicular on AC is to AN the perpendicular on BN as 2 to 3. And there will finally be this same ratio of 2 to 3 between the Sine of the angle EAD and the Sine of the angle FAN.

Hence one sees the reciprocal relation of the refractions of the ray on entering and on leaving one and the same transparent body: namely that if NA falling on the external surface AB is refracted into the direction AD, so the ray AD will be refracted on leaving the transparent body into the direction AN.

One sees also the reason for a noteworthy accident which happens in this refraction: which is this, that after a certain obliquity of the incident ray DA, it begins to be quite unable to penetrate into the other transparent substance. For if the angle DAQ or CBA is such that in the triangle ACB, CB is equal to $\frac{2}{3}$ of AB, or is greater, then AN cannot form one side of the triangle ANB, since it becomes equal to or greater than AB: so that the portion of wave BN cannot be found anywhere, neither consequently can AN, which ought to be perpendicular to it. And thus the incident ray DA does not then pierce the surface AB.

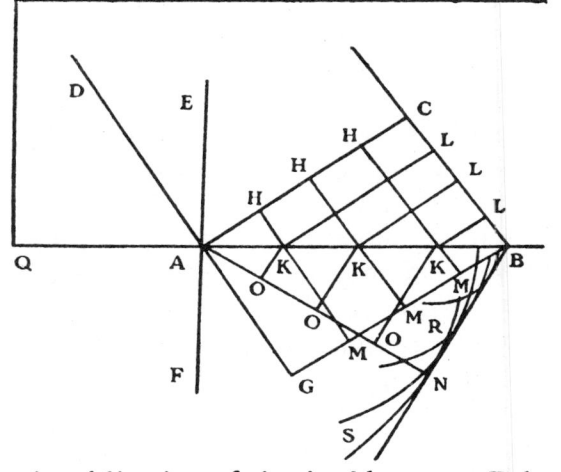

When

When the ratio of the velocities of the waves is as two to three, as in our example, which is that which obtains for glass and air, the angle DAQ must be more than 48 degrees 11 minutes in order that the ray DA may be able to pass by refraction. And when the ratio of the velocities is as 3 to 4, as it is very nearly in water and air, this angle DAQ must exceed 41 degrees 24 minutes. And this accords perfectly with experiment.

But it might here be asked: since the meeting of the wave AC against the surface AB ought to produce movement in the matter which is on the other side, why does no light pass there? To which the reply is easy if one remembers what has been said before. For although it generates an infinitude of partial waves in the matter which is at the other side of AB, these waves never have a common tangent line (either straight or curved) at the same moment; and so there is no line terminating the propagation of the wave AC beyond the plane AB, nor any place where the movement is gathered together in sufficiently great quantity to produce light. And one will easily see the truth of this, namely that CB being larger than $\frac{2}{3}$ of AB, the waves excited beyond the plane AB will have no common tangent if about the centres K one then draws circles having radii equal to $\frac{3}{2}$ of the lengths LB to which they correspond. For all these circles will be enclosed in one another and will all pass beyond the point B.

Now it is to be remarked that from the moment when the angle DAQ is smaller than is requisite to permit the refracted ray DA to pass into the other transparent substance, one finds that the interior reflexion which occurs at the surface AB is much augmented in brightness, as
is

ON LIGHT. CHAP. III

is easy to realize by experiment with a triangular prism; and for this our theory can afford this reason. When the angle DAQ is still large enough to enable the ray DA to pass, it is evident that the light from the portion AC of the wave is collected in a minimum space when it reaches BN. It appears also that the wave BN becomes so much the smaller as the angle CBA or DAQ is made less; until when the latter is diminished to the limit indicated a little previously, this wave BN is collected together always at one point. That is to say, that when the piece C of the wave AC has then arrived at B, the wave BN which is the propagation of AC is entirely reduced to the same point B. Similarly when the piece H has reached K, the part AH is entirely reduced to the same point K. This makes it evident that in proportion as the wave CA comes to meet the surface AB, there occurs a great quantity of movement along that surface; which movement ought also to spread within the transparent body and ought to have much re-enforced the partial waves which produce the interior reflexion against the surface AB, according to the laws of reflexion previously explained.

And because a slight diminution of the angle of incidence DAQ causes the wave BN, however great it was, to be reduced to zero, (for this angle being 49 degrees 11 minutes in the glass, the angle BAN is still 11 degrees 21 minutes, and the same angle being reduced by one degree only the angle BAN is reduced to zero, and so the wave BN reduced to a point) thence it comes about that the interior reflexion from being obscure becomes suddenly bright, so soon as the angle of incidence is such that it no longer gives passage to the refraction.

Now

Now as concerns ordinary external reflexion, that is to say which occurs when the angle of incidence DAQ is still large enough to enable the refracted ray to penetrate beyond the surface AB, this reflexion should occur against the particles of the substance which touches the transparent body on its outside. And it apparently occurs against the particles of the air or others mingled with the ethereal particles and larger than they. So on the other hand the external reflexion of these bodies occurs against the particles which compose them, and which are also larger than those of the ethereal matter, since the latter flows in their interstices. It is true that there remains here some difficulty in those experiments in which this interior reflexion occurs without the particles of air being able to contribute to it, as in vessels or tubes from which the air has been extracted.

Experience, moreover, teaches us that these two reflexions are of nearly equal force, and that in different transparent bodies they are so much the stronger as the refraction of these bodies is the greater. Thus one sees manifestly that the reflexion of glass is stronger than that of water, and that of diamond stronger than that of glass.

I will finish this theory of refraction by demonstrating a remarkable proposition which depends on it; namely, that a ray of light in order to go from one point to another, when these points are in different media, is refracted in such wise at the plane surface which joins these two media that it employs the least possible time: and exactly the same happens in the case of reflexion against a plane surface. Mr. Fermat was the first to propound this property of refraction, holding with us, and directly counter to the opinion of Mr. Des Cartes, that light passes

more

more slowly through glass and water than through air. But he assumed besides this a constant ratio of Sines, which we have just proved by these different degrees of velocity alone: or rather, what is equivalent, he assumed not only that the velocities were different but that the light took the least time possible for its passage, and thence deduced the constant ratio of the Sines. His demonstration, which may be seen in his printed works, and in the volume of letters of Mr. Des Cartes, is very long ; wherefore I give here another which is simpler and easier.

Let KF be the plane surface; A the point in the medium which the light traverses more easily, as the air; C the point in the other which is more difficult to penetrate, as water. And suppose that a ray has come from A, by B, to C, having been refracted at B according to the law demonstrated a little before; that is to say that, having drawn PBQ, which cuts the plane at right angles, let the sine of the angle ABP have to the sine of the angle CBQ the same ratio as the velocity of light in the medium where A is to the velocity of light in the medium where C is. It is to be shown that the time of passage of light along AB and BC taken together, is the shortest that can be. Let us assume that it may have come by other lines, and, in the first place, along AF, FC, so that

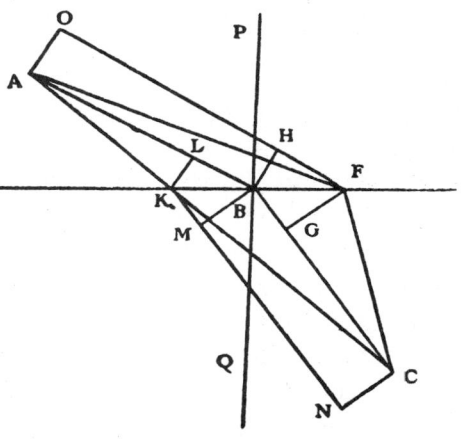

that the point of refraction F may be further from B·
than the point A; and let AO be a line perpendicular
to AB, and FO parallel to AB; BH perpendicular to
FO, and FG to BC.

Since then the angle HBF is equal to PBA, and the
angle BFG equal to QBC, it follows that the sine of the
angle HBF will also have the same ratio to the sine of
BFG, as the velocity of light in the medium A is to its
velocity in the medium C. But these sines are the straight
lines HF, BG, if we take BF as the semi-diameter of a
circle. Then these lines HF, BG, will bear to one another
the said ratio of the velocities. And, therefore, the time of
the light along HF, supposing that the ray had been OF,
would be equal to the time along BG in the interior of
the medium C. But the time along AB is equal to the
time along OH; therefore the time along OF is equal to
the time along AB, BG. Again the time along FC is
greater than that along GC; then the time along OFC will
be longer than that along ABC. But AF is longer than
OF, then the time along AFC will by just so much
more exceed the time along ABC.

Now let us assume that the ray has come from A to C
along AK, KC; the point of refraction K being nearer
to A than the point B is; and let CN be the perpendicular upon BC, KN parallel to BC: BM perpendicular
upon KN, and KL upon BA.

Here BL and KM are the sines of angles BKL, KBM;
that is to say, of the angles PBA, QBC; and therefore
they are to one another as the velocity of light in the
medium A is to the velocity in the medium C. Then
the time along LB is equal to the time along KM; and
since the time along BC is equal to the time along MN, the
time

ON LIGHT. Chap. IV

time along LBC will be equal to the time along KMN. But the time along AK is longer than that along AL: hence the time along AKN is longer than that along ABC. And KC being longer than KN, the time along AKC will exceed, by as much more, the time along ABC. Hence it appears that the time along ABC is the shortest possible; which was to be proven.

CHAPTER IV
On the Refraction of the Air

We have shown how the movement which constitutes light spreads by spherical waves in any homogeneous matter. And it is evident that when the matter is not homogeneous, but of such a constitution that the movement is communicated in it more rapidly toward one side than toward another, these waves cannot be spherical: but that they must acquire their figure according to the different distances over which the successive movement passes in equal times.

It is thus that we shall in the first place explain the refractions which occur in the air, which extends from here to the clouds and beyond. The effects of which refractions are very remarkable; for by them we often see objects which the rotundity of the Earth ought otherwise to hide; such as Islands, and the tops of mountains when one is at sea. Because also of them the Sun and the Moon appear as risen before in fact they have, and appear to set later:

later: so that at times the Moon has been seen eclipsed while the Sun appeared still above the horizon. And so also the heights of the Sun and of the Moon, and those of all the Stars always appear a little greater than they are in reality, because of these same refractions, as Astronomers know. But there is one experiment which renders this refraction very evident; which is that of fixing a telescope on some spot so that it views an object, such as a steeple or a house, at a distance of half a league or more. If then you look through it at different hours of the day, leaving it always fixed in the same way, you will see that the same spots of the object will not always appear at the middle of the aperture of the telescope, but that generally in the morning and in the evening, when there are more vapours near the Earth, these objects seem to rise higher, so that the half or more of them will no longer be visible; and so that they seem lower toward mid-day when these vapours are dissipated.

Those who consider refraction to occur only in the surfaces which separate transparent bodies of different nature, would find it difficult to give a reason for all that I have just related; but according to our Theory the thing is quite easy. It is known that the air which surrounds us, besides the particles which are proper to it and which float in the ethereal matter as has been explained, is full also of particles of water which are raised by the action of heat; and it has been ascertained further by some very definite experiments that as one mounts up higher the density of air diminishes in proportion. Now whether the particles of water and those of air take part, by means of the particles of ethereal matter, in the movement which constitutes light, but have a less prompt recoil than these,
or

or whether the encounter and hindrance which these particles of air and water offer to the propagation of movement of the ethereal progress, retard the progression, it follows that both kinds of particles flying amidst the ethereal particles, must render the air, from a great height down to the Earth, gradually less easy for the spreading of the waves of light.

Whence the configuration of the waves ought to become nearly such as this figure represents: namely, if A is a light, or the visible point of a steeple, the waves which start from it ought to spread more widely upwards and less widely downwards, but in other directions more or less as they approximate to these two extremes. This being so, it necessarily follows that every line intersecting one of these waves at right angles will pass above the point A, always excepting the one line which is perpendicular to the horizon.

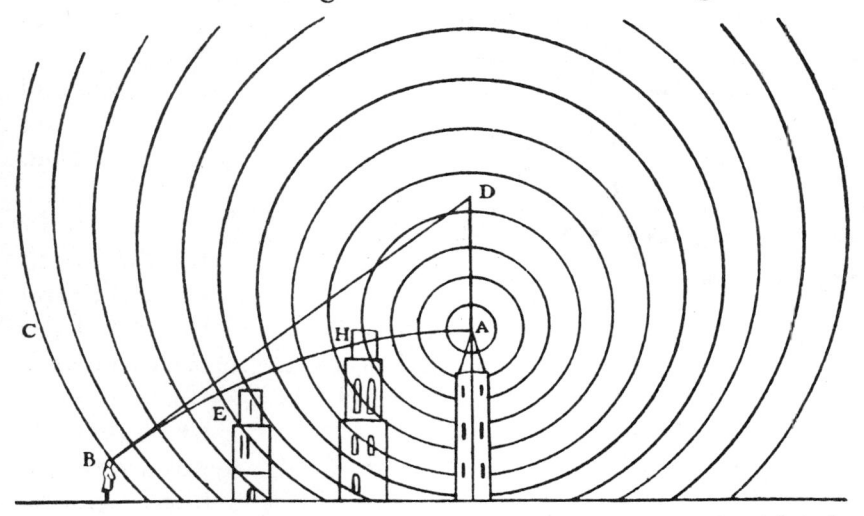

Let

Let BC be the wave which brings the light to the spectator who is at B, and let BD be the straight line which intersects this wave at right angles. Now because the ray or straight line by which we judge the spot where the object appears to us is nothing else than the perpendicular to the wave that reaches our eye, as will be understood by what was said above, it is manifest that the point A will be perceived as being in the line BD, and therefore higher than in fact it is.

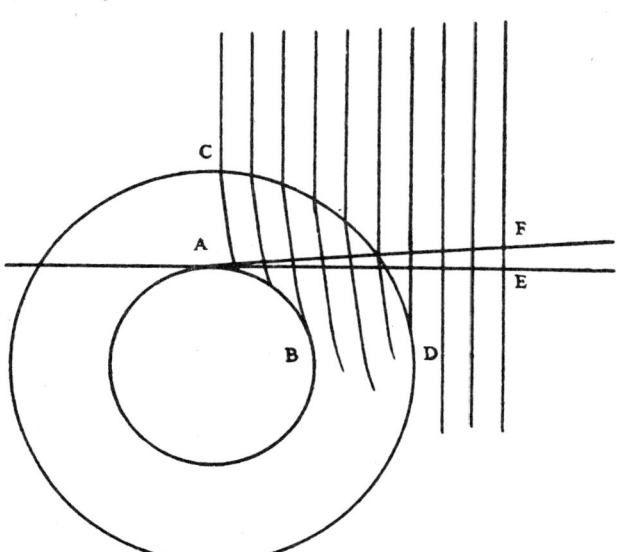

Similarly if the Earth be AB, and the top of the Atmosphere CD, which probably is not a well defined spherical surface (since we know that the air becomes rare in proportion as one ascends, for above there is so much less of it to press down upon it), the waves of light from the sun coming, for instance, in such a way that so long as they have not reached the Atmosphere CD the straight line AE intersects them perpendicularly, they ought, when they enter the Atmosphere, to advance more quickly in elevated regions than in regions nearer to the Earth. So that if

ON LIGHT. Chap. IV

CA is the wave which brings the light to the spectator at A, its region C will be the furthest advanced; and the straight line AF, which intersects this wave at right angles, and which determines the apparent place of the Sun, will pass above the real Sun, which will be seen along the line AE. And so it may occur that when it ought not to be visible in the absence of vapours, because the line AE encounters the rotundity of the Earth, it will be perceived in the line AF by refraction. But this angle EAF is scarcely ever more than half a degree because the attenuation of the vapours alters the waves of light but little. Furthermore these refractions are not altogether constant in all weathers, particularly at small elevations of 2 or 3 degrees; which results from the different quantity of aqueous vapours rising above the Earth.

And this same thing is the cause why at certain times a distant object will be hidden behind another less distant one, and yet may at another time be able to be seen, although the spot from which it is viewed is always the same. But the reason for this effect will be still more evident from what we are going to remark touching the curvature of rays. It appears from the things explained above that the progression or propagation of a small part of a wave of light is properly what one calls a ray. Now these rays, instead of being straight as they are in homogeneous media, ought to be curved in an atmosphere of unequal penetrability. For they necessarily follow from the object to the eye the line which intersects at right angles all the progressions of the waves, as in the first figure the line AEB does, as will be shown hereafter; and it is this line which determines what interposed bodies would or would not hinder us from seeing the object. For

although

although the point of the steeple A appears raised to D, it would yet not appear to the eye B if the tower H was between the two, because it crosses the curve AEB. But the tower E, which is beneath this curve, does not hinder the point A from being seen. Now according as the air near the Earth exceeds in density that which is higher, the curvature of the ray AEB becomes greater: so that at certain times it passes above the summit E, which allows the point A to be perceived by the eye at B; and at other times it is intercepted by the same tower E which hides A from this same eye.

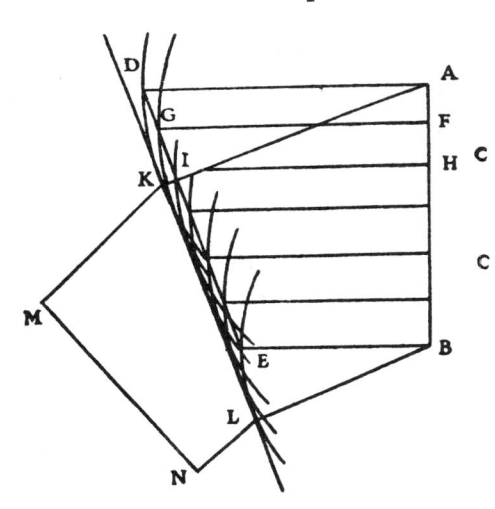

But to demonstrate this curvature of the rays conformably to all our preceding Theory, let us imagine that AB is a small portion of a wave of light coming from the side C, which we may consider as a straight line. Let us also suppose that it is perpendicular to the Horizon, the portion B being nearer to the Earth than the portion A; and that because the vapours are less hindering at A than at B, the particular wave which comes from the point A spreads through a certain space AD while the particular wave which starts from the point B spreads through a shorter space BE; AD and BE being parallel to the Horizon. Further, supposing the straight lines FG, HI, etc., to be
drawn

ON LIGHT. CHAP. IV 51

drawn from an infinitude of points in the straight line AB and to terminate on the line DE (which is straight or may be considered as such), let the different penetrabilities at the different heights in the air between A and B be represented by all these lines; so that the particular wave, originating from the point F, will spread across the space FG, and that from the point H across the space HI, while that from the point A spreads across the space AD.

Now if about the centres A, B, one describes the circles DK, EL, which represent the spreading of the waves which originate from these two points, and if one draws the straight line KL which touches these two circles, it is easy to see that this same line will be the common tangent to all the other circles drawn about the centres F, H, etc.; and that all the points of contact will fall within that part of this line which is comprised between the perpendiculars AK, BL. Then it will be the line KL which will terminate the movement of the particular waves originating from the points of the wave AB; and this movement will be stronger between the points KL, than anywhere else at the same instant, since an infinitude of circumferences concur to form this straight line; and consequently KL will be the propagation of the portion of wave AB, as has been said in explaining reflexion and ordinary refraction. Now it appears that AK and BL dip down toward the side where the air is less easy to penetrate: for AK being longer than BL, and parallel to it, it follows that the lines AB and KL, being prolonged, would meet at the side L. But the angle K is a right angle: hence KAB is necessarily acute, and consequently less than DAB. If one investigates in the same way the progression of the portion of the wave KL, one will find that after a further time it has arrived

arrived at MN in such a manner that the perpendiculars KM, LN, dip down even more than do AK, BL. And this suffices to show that the ray will continue along the curved line which intersects all the waves at right angles, as has been said.

CHAPTER V

On the Strange Refraction of Iceland Crystal

I

THERE is brought from Iceland, which is an Island in the North Sea, in the latitude of 66 degrees, a kind of Crystal or transparent stone, very remarkable for its figure and other qualities, but above all for its strange refractions. The causes of this have seemed to me to be worthy of being carefully investigated, the more so because amongst transparent bodies this one alone does not follow the ordinary rules with respect to rays of light. I have even been under some necessity to make this research, because the refractions of this Crystal seemed to overturn our preceding explanation of regular refraction; which explanation, on the contrary, they strongly confirm, as will be seen after they have been brought under the same principle. In Iceland are found great lumps of this Crystal, some of which I have seen of 4 or 5 pounds. But it occurs also in other countries, for I have had some of the same sort which had been found in France near the town of Troyes in Champagne, and some others which came from the Island of Corsica, though both were
less

ON LIGHT. Chap. V

less clear and only in little bits, scarcely capable of letting any effect of refraction be observed.

2. The first knowledge which the public has had about it is due to Mr. Erasmus Bartholinus, who has given a description of Iceland Crystal and of its chief phenomena. But here I shall not desist from giving my own, both for the instruction of those who may not have seen his book, and because as respects some of these phenomena there is a slight difference between his observations and those which I have made: for I have applied myself with great exactitude to examine these properties of refraction, in order to be quite sure before undertaking to explain the causes of them.

3. As regards the hardness of this stone, and the property which it has of being easily split, it must be considered rather as a species of Talc than of Crystal. For an iron spike effects an entrance into it as easily as into any other Talc or Alabaster, to which it is equal in gravity.

4. The pieces of it which are found have the figure of an oblique parallelopiped; each of the six faces being a parallelogram; and it admits of being split in three directions parallel to two of these opposed faces. Even in such wise, if you will, that all the six faces are equal and similar rhombuses. The figure here added represents a piece of this Crystal. The obtuse angles of all the parallelograms, as C, D, here, are angles of 101 degrees 52 minutes, and

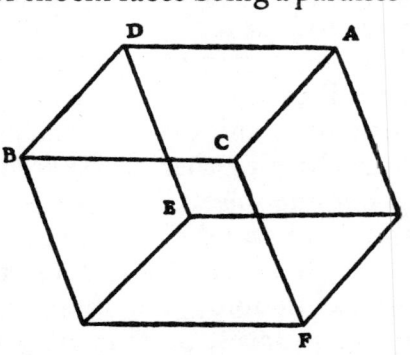

and consequently the acute angles, such as A and B, are of 78 degrees 8 minutes.

5. Of the solid angles there are two opposite to one another, such as C and E, which are each composed of three equal obtuse plane angles. The other six are composed of two acute angles and one obtuse. All that I have just said has been likewise remarked by Mr. Bartholinus in the aforesaid treatise; if we differ it is only slightly about the values of the angles. He recounts moreover some other properties of this Crystal; to wit, that when rubbed against cloth it attracts straws and other light things as do amber, diamond, glass, and Spanish wax. Let a piece be covered with water for a day or more, the surface loses its natural polish. When aquafortis is poured on it it produces ebullition, especially, as I have found, if the Crystal has been pulverized. I have also found by experiment that it may be heated to redness in the fire without being in anywise altered or rendered less transparent; but a very violent fire calcines it nevertheless. Its transparency is scarcely less than that of water or of Rock Crystal, and devoid of colour. But rays of light pass through it in another fashion and produce those marvellous refractions the causes of which I am now going to try to explain; reserving for the end of this Treatise the statement of my conjectures touching the formation and extraordinary configuration of this Crystal.

6. In all other transparent bodies that we know there is but one sole and simple refraction; but in this substance there are two different ones. The effect is that objects seen through it, especially such as are placed right against it, appear double; and that a ray of sunlight, falling on one of its surfaces, parts itself into two rays and traverses the Crystal thus.

7. It

7. It is again a general law in all other transparent bodies that the ray which falls perpendicularly on their surface passes straight on without suffering refraction, and that an oblique ray is always refracted. But in this Crystal the perpendicular ray suffers refraction, and there are oblique rays which pass through it quite straight.

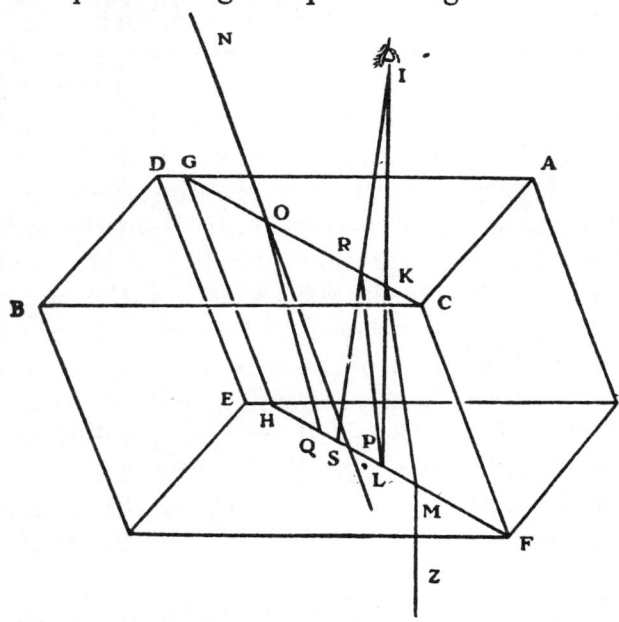

8. But in order to explain these phenomena more particularly, let there be, in the first place, a piece ABFE of the same Crystal, and let the obtuse angle ACB, one of the three which constitute the equilateral solid angle C, be divided into two equal parts by the straight line CG, and let it be conceived that the Crystal is intersected by a plane which passes through this line and through the side CF, which plane will necessarily be perpendicular to the

the surface AB; and its section in the Crystal will form a parallelogram GCFH. We will call this section the principal section of the Crystal.

9. Now if one covers the surface AB, leaving there only a small aperture at the point K, situated in the straight line CG, and if one exposes it to the sun, so that his rays face it perpendicularly above, then the ray IK will divide itself at the point K into two, one of which will continue to go on straight by KL, and the other will separate itself along the straight line KM, which is in the plane GCFH, and which makes with KL an angle of about 6 degrees 40 minutes, tending from the side of the solid angle C; and on emerging from the other side of the Crystal it will turn again parallel to IK, along MZ. And as, in this extraordinary refraction, the point M is seen by the refracted ray MKI, which I consider as going to the eye at I, it necessarily follows that the point L, by virtue of the same refraction, will be seen by the refracted ray LRI, so that LR will be parallel to MK if the distance from the eye KI is supposed very great. The point L appears then as being in the straight line IRS; but the same point appears also, by ordinary refraction, to be in the straight line IK, hence it is necessarily judged to be double. And similarly if L be a small hole in a sheet of paper or other substance which is laid against the Crystal, it will appear when turned towards daylight as if there were two holes, which will seem the wider apart from one another the greater the thickness of the Crystal.

10. Again, if one turns the Crystal in such wise that an incident ray NO, of sunlight, which I suppose to be in the plane continued from GCFH, makes with GC an angle

angle of 73 degrees and 20 minutes, and is consequently nearly parallel to the edge CF, which makes with FH an angle of 70 degrees 57 minutes, according to the calculation which I shall put at the end, it will divide itself at the point O into two rays, one of which will continue along OP in a straight line with NO, and will similarly pass out of the other side of the crystal without any refraction; but the other will be refracted and will go along OQ. And it must be noted that it is special to the plane through GCF and to those which are parallel to it, that all incident rays which are in one of these planes continue to be in it after they have entered the Crystal and have become double; for it is quite otherwise for rays in all other planes which intersect the Crystal, as we shall see afterwards.

11. I recognized at first by these experiments and by some others that of the two refractions which the ray suffers in this Crystal, there is one which follows the ordinary rules; and it is this to which the rays KL and OQ belong. This is why I have distinguished this ordinary refraction from the other; and having measured it by exact observation, I found that its proportion, considered as to the Sines of the angles which the incident and refracted rays make with the perpendicular, was very precisely that of 5 to 3, as was found also by Mr. Bartholinus, and consequently much greater than that of Rock Crystal, or of glass, which is nearly 3 to 2.

12. The mode of making these observations exactly is as follows. Upon a leaf of paper fixed on a thoroughly flat table there is traced a black line AB, and two others, CED and KML, which cut it at right angles and are more or less distant from one another according

as it is desired to examine a ray that is more or less oblique. Then place the Crystal upon the intersection E so that the line AB concurs with that which bisects the obtuse angle of the lower surface, or with some line parallel to it. Then by placing the eye directly above the line AB it will appear single only; and one will see that the portion viewed through the Crystal

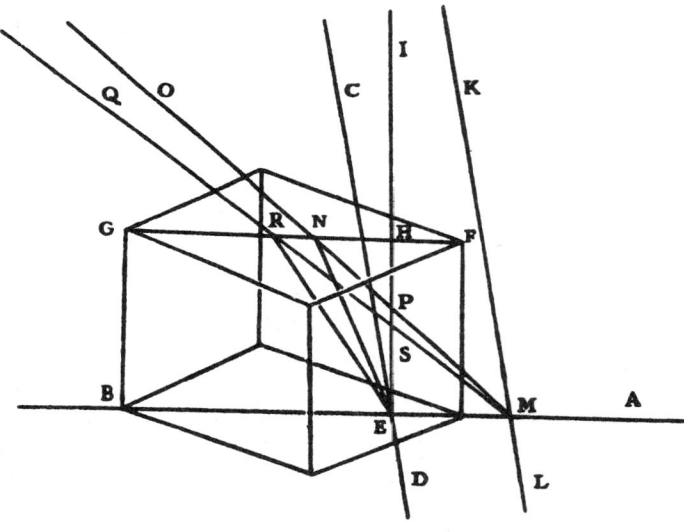

and the portions which appear outside it, meet together in a straight line : but the line CD will appear double, and one can distinguish the image which is due to regular refraction by the circumstance that when one views it with both eyes it seems raised up more than the other, or again by the circumstance that, when the Crystal is turned around on the paper, this image remains stationary, whereas the other image shifts and moves entirely around. Afterwards let the eye be placed at I (remaining
always

always in the plane perpendicular through AB) so that it views the image which is formed by regular refraction of the line CD making a straight line with the remainder of that line which is outside the Crystal. And then, marking on the surface of the Crystal the point H where the intersection E appears, this point will be directly above E. Then draw back the eye towards O, keeping always in the plane perpendicular through AB, so that the image of the line CD, which is formed by ordinary refraction, may appear in a straight line with the line KL viewed without refraction; and then mark on the Crystal the point N where the point of intersection E appears.

13. Then one will know the length and position of the lines NH, EM, and of HE, which is the thickness of the Crystal: which lines being traced separately upon a plan, and then joining NE and NM which cuts HE at P, the proportion of the refraction will be that of EN to NP, because these lines are to one another as the sines of the angles NPH, NEP, which are equal to those which the incident ray ON and its refraction NE make with the perpendicular to the surface. This proportion, as I have said, is sufficiently precisely as 5 to 3, and is always the same for all inclinations of the incident ray.

14. The same mode of observation has also served me for examining the extraordinary or irregular refraction of this Crystal. For, the point H having been found and marked, as aforesaid, directly above the point E, I observed the appearance of the line CD, which is made by the extraordinary refraction; and having placed the eye at Q, so that this appearance made a straight line with the line KL viewed without refraction, I ascertained the triangles REH, RES, and consequently the angles RSH, RES,

RES, which the incident and the refracted ray make with the perpendicular.

15. But I found in this refraction that the ratio of FR to RS was not constant, like the ordinary refraction, but that it varied with the varying obliquity of the incident ray.

16. I found also that when QRE made a straight line, that is, when the incident ray entered the Crystal without being refracted (as I ascertained by the circumstance that then the point E viewed by the extraordinary refraction appeared in the line CD, as seen without refraction) I found, I say, then that the angle QRG was 73 degrees 20 minutes, as has been already remarked; and so it is not the ray parallel to the edge of the Crystal, which crosses it in a straight line without being refracted, as Mr. Bartholinus believed, since that inclination is only 70 degrees 57 minutes, as was stated above. And this is to be noted, in order that no one may search in vain for the cause of the singular property of this ray in its parallelism to the edges mentioned.

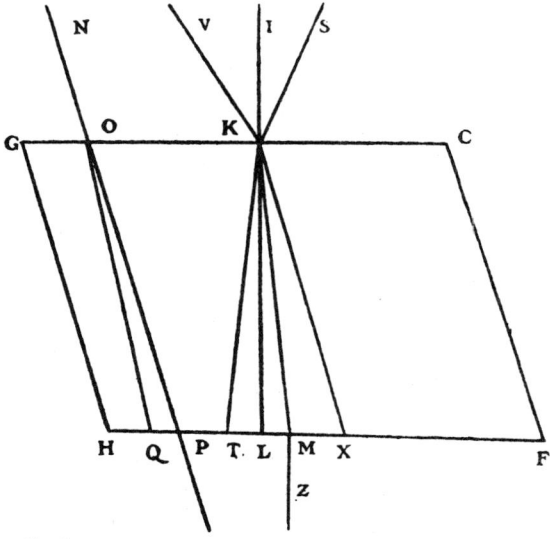

17. Finally, continuing my observations to discover the nature

nature of this refraction, I learned that it obeyed the following remarkable rule. Let the parallelogram GCFH, made by the principal section of the Crystal, as previously determined, be traced separately. I found then that always, when the inclinations of two rays which come from opposite sides, as VK, SK here, are equal, their refractions KX and KT meet the bottom line HF in such wise that points X and T are equally distant from the point M, where the refraction of the perpendicular ray IK falls; and this occurs also for refractions in other sections of this Crystal. But before speaking of those, which have also other particular properties, we will investigate the causes of the phenomena which I have already reported.

It was after having explained the refraction of ordinary transparent bodies by means of the spherical emanations of light, as above, that I resumed my examination of the nature of this Crystal, wherein I had previously been unable to discover anything.

18. As there were two different refractions, I conceived that there were also two different emanations of waves of light, and that one could occur in the ethereal matter extending through the body of the Crystal. Which matter, being present in much larger quantity than is that of the particles which compose it, was alone capable of causing transparency, according to what has been explained heretofore. I attributed to this emanation of waves the regular refraction which is observed in this stone, by supposing these waves to be ordinarily of spherical form, and having a slower progression within the Crystal than they have outside it; whence proceeds refraction as I have demonstrated.

19. As to the other emanation which should produce the

the irregular refraction, I wished to try what Elliptical waves, or rather spheroidal waves, would do; and these I supposed would spread indifferently both in the ethereal matter diffused throughout the crystal and in the particles of which it is composed, according to the last mode in which I have explained transparency. It seemed to me that the disposition or regular arrangement of these particles could contribute to form spheroidal waves (nothing more being required for this than that the successive movement of light should spread a little more quickly in one direction than in the other) and I scarcely doubted that there were in this crystal such an arrangement of equal and similar particles, because of its figure and of its angles with their determinate and invariable measure. Touching which particles, and their form and disposition, I shall, at the end of this Treatise, propound my conjectures and some experiments which confirm them.

20. The double emission of waves of light, which I had imagined, became more probable to me after I had observed a certain phenomenon in the ordinary [Rock] Crystal, which occurs in hexagonal form, and which, because of this regularity, seems also to be composed of particles, of definite figure, and ranged in order. This was, that this crystal, as well as that from Iceland, has a double refraction, though less evident. For having had cut from it some well polished Prisms of different sections, I remarked in all, in viewing through them the flame of a candle or the lead of window panes, that everything appeared double, though with images not very distant from one another. Whence I understood the reason why this substance, though so transparent, is useless for Telescopes, when they have ever so little length.

21. Now

ON LIGHT. Chap. V

21. Now this double refraction, according to my Theory hereinbefore established, seemed to demand a double emission of waves of light, both of them spherical (for both the refractions are regular) and those of one series a little slower only than the others. For thus the phenomenon is quite naturally explained, by postulating substances which serve as vehicle for these waves, as I have done in the case of Iceland Crystal. I had then less trouble after that in admitting two emissions of waves in one and the same body. And since it might have been objected that in composing these two kinds of crystal of equal particles of a certain figure, regularly piled, the interstices which these particles leave and which contain the ethereal matter would scarcely suffice to transmit the waves of light which I have localized there, I removed this difficulty by regarding these particles as being of a very rare texture, or rather as composed of other much smaller particles, between which the ethereal matter passes quite freely. This, moreover, necessarily follows from that which has been already demonstrated touching the small quantity of matter of which the bodies are built up.

22. Supposing then these spheroidal waves besides the spherical ones, I began to examine whether they could serve to explain the phenomena of the irregular refraction, and how by these same phenomena I could determine the figure and position of the spheroids: as to which I obtained at last the desired success, by proceeding as follows.

23. I considered first the effect of waves so formed, as respects the ray which falls perpendicularly on the flat surface of a transparent body in which they should spread in this manner. I took AB for the exposed region of the surface. And, since a ray perpendicular to a plane, and
coming

coming from a very distant source of light, is nothing else, according to the precedent Theory, than the incidence of a portion of the wave parallel to that plane, I supposed the straight line RC, parallel and equal to AB, to be a portion of a wave of light, in which an infinitude of points such as RH*h*C come to meet the surface AB at the points AK*k*B. Then instead of the hemispherical partial waves which in a body of ordinary refraction would spread from each of these last points, as we have above explained in treating of refraction, these must here be hemispheroids. The axes (or rather the major diameters) of these I supposed to be oblique to the plane AB, as is AV the semi-axis or semi-major diameter of the spheroid SVT, which represents the partial wave coming from the point A, after the wave RC has reached AB. I say axis or major diameter, because the same ellipse SVT may be considered as the section of a spheroid of which the axis is AZ perpendicular to AV. But, for the present, without yet deciding one or other, we will consider these spheroids only in those sections of them which make ellipses in the plane of this figure. Now taking a certain space of time during which the wave SVT has spread from A, it would needs be that from all the other points K*k*B there should proceed, in the same time, waves similar to SVT and similarly situated. And the common tangent NQ of all these semi-ellipses would be the propagation of the wave RC which fell on AB, and
would

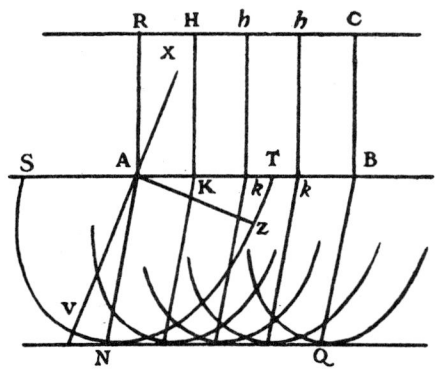

ON LIGHT. Chap. V

would be the place where this movement occurs in much greater amount than anywhere else, being made up of arcs of an infinity of ellipses, the centres of which are along the line AB.

24. Now it appeared that this common tangent NQ was parallel to AB, and of the same length, but that it was not directly opposite to it, since it was comprised between the lines AN, BQ, which are diameters of ellipses having A and B for centres, conjugate with respect to diameters which are not in the straight line AB. And in this way I comprehended, a matter which had seemed to me very difficult, how a ray perpendicular to a surface could suffer refraction on entering a transparent body; seeing that the wave RC, having come to the aperture AB, went on forward thence, spreading between the parallel lines AN, BQ, yet itself remaining always parallel to AB, so that here the light does not spread along lines perpendicular to its waves, as in ordinary refraction, but along lines cutting the waves obliquely.

25. Inquiring subsequently what might be the position and form of these spheroids in the crystal, I considered that all the six faces produced precisely the same refractions. Taking, then, the parallelopiped AFB, of which the obtuse solid angle C is contained between the three equal plane angles, and imagining in it the three principal sections, one of which is perpendicular to the face DC and passes through the edge CF, another perpendicular to the face BF passing through the edge CA,

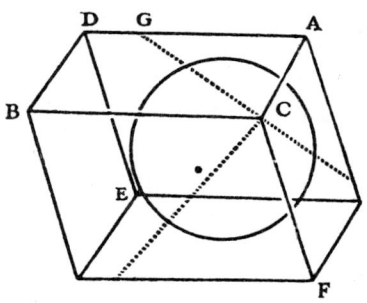

CA, and the third perpendicular to the face AF passing through the edge BC; I knew that the refractions of the incident rays belonging to these three planes were all similar. But there could be no position of the spheroid which would have the same relation to these three sections except that in which the axis was also the axis of the solid angle C. Consequently I saw that the axis of this angle, that is to say the straight line which traversed the crystal from the point C with equal inclination to the edges CF, CA, CB was the line which determined the position of the axis of all the spheroidal waves which one imagined to originate from some point, taken within or on the surface of the crystal, since all these spheroids ought to be alike, and have their axes parallel to one another.

26. Considering after this the plane of one of these three sections, namely that through GCF, the angle of which is 109 degrees 3 minutes, since the angle F was shown above to be 70 degrees 57 minutes; and, imagining a spheroidal wave about the centre C, I knew, because I have just explained it, that its axis must be in the same plane, the half of which axis I have marked CS in the next figure: and seeking by calculation (which will be given with others at the end of this discourse) the value of the angle CGS, I found it 45 degrees 20 minutes.

27. To know from this the form of this spheroid, that is to say the proportion of the semi-diameters CS, CP, of its elliptical section, which are perpendicular to one another, I considered that the point M where the ellipse is touched by the straight line FH, parallel to CG, ought to be so situated that CM makes with the perpendicular CL an angle of 6 degrees 40 minutes; since, this being so, this ellipse satisfies what has been said about the refraction of the

the ray perpendicular to the surface CG, which is inclined to the perpendicular CL by the same angle. This, then, being thus disposed, and taking CM at 100,000 parts, I found by the calculation which will be given at the end, the semi-major diameter CP to be 105,032, and the semi-axis CS to be 93,410, the ratio of which numbers is very nearly 9 to 8; so that the spheroid was of the kind which resembles a compressed sphere, being generated by the revolution of an ellipse about its smaller diameter. I found also the value of CG the semi-diameter parallel to the tangent ML to be 98,779.

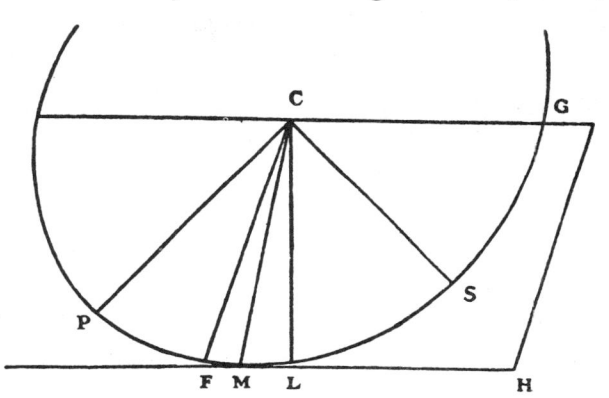

28. Now passing to the investigation of the refractions which obliquely incident rays must undergo, according to our hypothesis of spheroidal waves, I saw that these refractions depended on the ratio between the velocity of movement of the light outside the crystal in the ether, and that within the crystal. For supposing, for example, this proportion to be such that while the light in the crystal forms the spheroid GSP, as I have just said, it forms outside a sphere the semi-diameter of which is equal to the line N which will be determined hereafter, the following is the way of finding the refraction of the incident rays. Let there be such a ray RC falling upon the surface

surface CK. Make CO perpendicular to RC, and across the angle KCO adjust OK, equal to N and perpendicular to CO; then draw KI, which touches the Ellipse GSP, and from the point of contact I join IC, which will be the required refraction of the ray RC. The demonstration of this is, it will be seen, entirely similar to that of which we made use in explaining ordinary refraction. For the

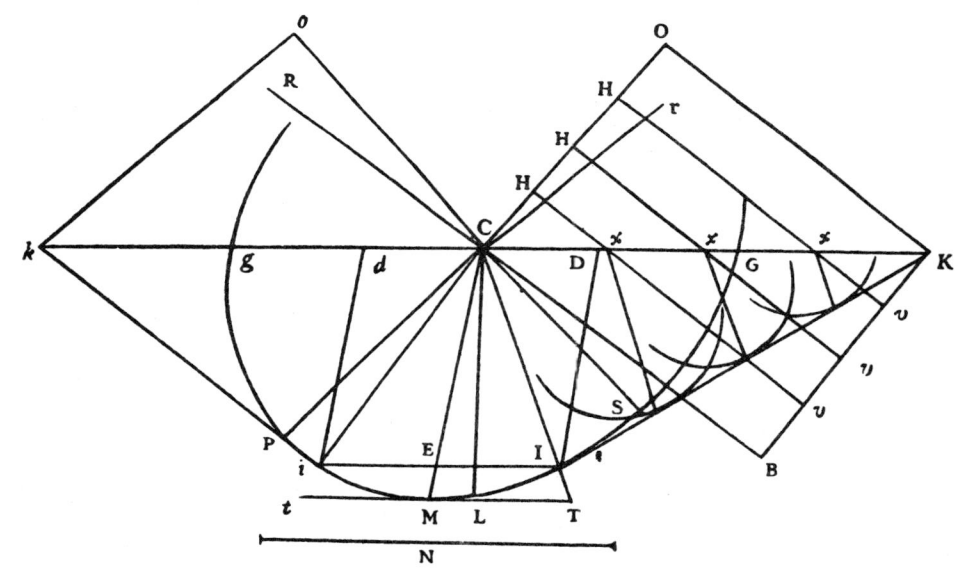

refraction of the ray RC is nothing else than the progression of the portion C of the wave CO, continued in the crystal. Now the portions H of this wave, during the time that O came to K, will have arrived at the surface CK along the straight lines H*x*, and will moreover have produced in the crystal around the centres *x* some hemi-spheroidal partial waves similar to the hemi-spheroidal GSP*g*, and similarly disposed, and of which the major and

ON LIGHT. Chap. V

and minor diameters will bear the same proportions to the lines *xv* (the continuations of the lines H*x* up to KB parallel to CO) that the diameters of the spheroid GSP*g* bear to the line CB, or N. And it is quite easy to see that the common tangent of all these spheroids, which are here represented by Ellipses, will be the straight line IK, which consequently will be the propagation of the wave CO; and the point I will be that of the point C, conformably with that which has been demonstrated in ordinary refraction.

Now as to finding the point of contact I, it is known that one must find CD a third proportional to the lines CK, CG, and draw DI parallel to CM, previously determined, which is the conjugate diameter to CG; for then, by drawing KI it touches the Ellipse at I.

29. Now as we have found CI the refraction of the ray RC, similarly one will find C*i* the refraction of the ray *r*C, which comes from the opposite side, by making C*o* perpendicular to *r*C and following out the rest of the construction as before. Whence one sees that if the ray *r*C is inclined equally with RC, the line C*d* will necessarily be equal to CD, because C*k* is equal to CK, and C*g* to CG. And in consequence I*i* will be cut at E into equal parts by the line CM, to which DI and *di* are parallel. And because CM is the conjugate diameter to CG, it follows that *i*I will be parallel to *g*G. Therefore if one prolongs the refracted rays CI, C*i*, until they meet the tangent ML at T and *t*, the distances MT, M*t*, will also be equal. And so, by our hypothesis, we explain perfectly the phenomenon mentioned above; to wit, that when there are two rays equally inclined, but coming from opposite sides, as here the rays RC, *rc*, their refractions diverge equally from the line
followed

followed by the refraction of the ray perpendicular to the surface, by considering these divergences in the direction parallel to the surface of the crystal.

30. To find the length of the line N, in proportion to CP, CS, CG, it must be determined by observations of the irregular refraction which occurs in this section of the crystal; and I find thus that the ratio of N to GC is just a little less than 8 to 5. And having regard to some other observations and phenomena of which I shall speak afterwards, I put N at 156,962 parts, of which the semi-diameter CG is found to contain 98,779, making this ratio 8 to $5\frac{1}{29}$. Now this proportion, which there is between the line N and CG, may be called the Proportion of the Refraction; similarly as in glass that of 3 to 2, as will be manifest when I shall have explained a short process in the preceding way to find the irregular refractions.

31. Supposing then, in the next figure, as previously, the surface of the crystal gG, the Ellipse GPg, and the line N; and CM the refraction of the perpendicular ray FC, from which it diverges by 6 degrees 40 minutes. Now let there be some other ray RC, the refraction of which must be found.

About the centre C, with semi-diameter CG, let the circumference gRG be described, cutting the ray RC at R; and let RV be the perpendicular on CG. Then as the line N is to CG let CV be to CD, and let DI be drawn parallel to CM, cutting the Ellipse gMG at I; then joining CI, this will be the required refraction of the ray RC. Which is demonstrated thus.

Let CO be perpendicular to CR, and across the angle OCG let OK be adjusted, equal to N and perpendicular to CO, and let there be drawn the straight line KI, which if it

is

ON LIGHT. Chap. V

is demonstrated to be a tangent to the Ellipse at I, it will be evident by the things heretofore explained that CI is the refraction of the ray RC. Now since the angle RCO is a right angle, it is easy to see that the right-angled triangles RCV, KCO, are similar. As then, CK is to KO, so also

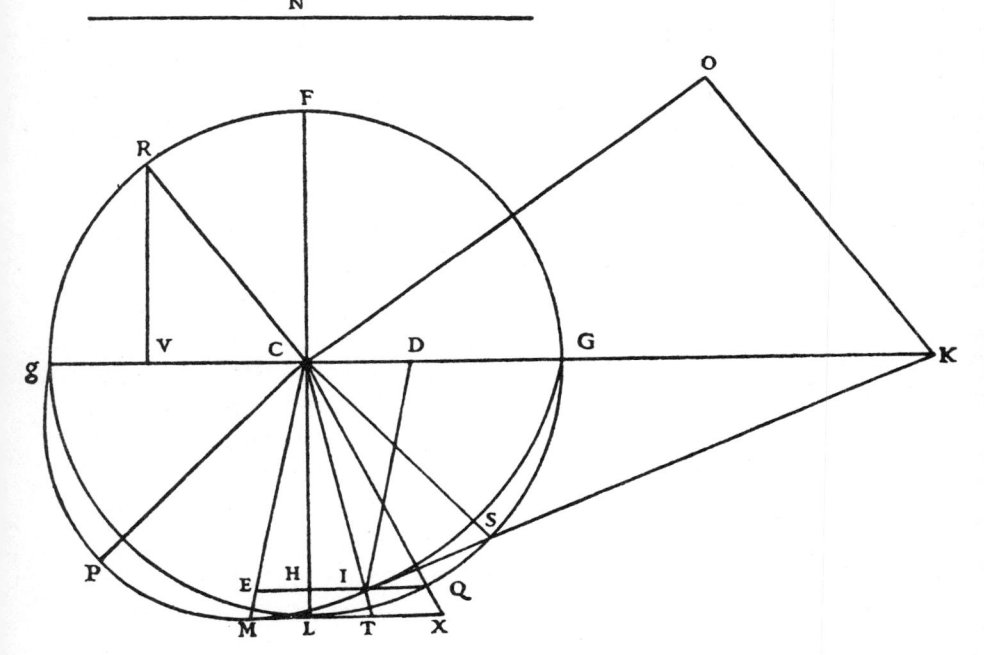

is RC to CV. But KO is equal to N, and RC to CG: then as CK is to N so will CG be to CV. But as N is to CG, so, by construction, is CV to CD. Then as CK is to CG so is CG to CD. And because DI is parallel to CM, the conjugate diameter to CG, it follows that KI touches the Ellipse at I; which remained to be shown.

32. One sees then that as there is in the refraction of
ordinary

ordinary media a certain constant proportion between the sines of the angles which the incident ray and the refracted ray make with the perpendicular, so here there is such a proportion between CV and CD or IE; that is to say between the Sine of the angle which the incident ray makes with the perpendicular, and the horizontal intercept, in the Ellipse, between the refraction of this ray and the diameter CM. For the ratio of CV to CD is, as has been said, the same as that of N to the semi-diameter CG.

33. I will add here, before passing away, that in comparing together the regular and irregular refraction of this crystal, there is this remarkable fact, that if ABPS be the spheroid by which light spreads in the Crystal in a certain space of time (which spreading, as has been said, serves for the irregular refraction), then the inscribed sphere BVST is the extension in the same space of time of the light which serves for the regular refraction.

For we have stated before this, that the line N being the radius of a spherical wave of light in air, while in the crystal it spread through the spheroid ABPS, the ratio of N to CS will be 156,962 to 93,410. But it has also been stated that the proportion of the regular refraction was 5 to 3; that is to say, that N being the radius of a spherical wave of light in air, its extension in the crystal would, in the same space of time, form a sphere the radius of which would be to N as 3 to 5. Now 156,962 is to 93,410 as 5 to 3 less $\frac{1}{41}$. So that it is sufficiently nearly, and may be

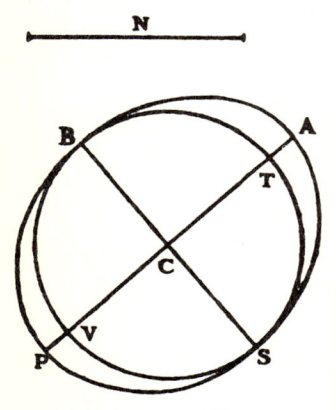

ON LIGHT. Chap. V

be exactly, the sphere BVST, which the light describes for the regular refraction in the crystal, while it describes the spheroid BPSA for the irregular refraction, and while it describes the sphere of radius N in air outside the crystal.

Although then there are, according to what we have supposed, two different propagations of light within the crystal, it appears that it is only in directions perpendicular to the axis BS of the spheroid that one of these propagations occurs more rapidly than the other; but that they have an equal velocity in the other direction, namely, in that parallel to the same axis BS, which is also the axis of the obtuse angle of the crystal.

34. The proportion of the refraction being what we have just seen, I will now show that there necessarily follows thence that notable property of the ray which falling obliquely on the surface of the crystal enters it without suffering refraction. For supposing the same things as before, and that the ray RC makes with the same surface *g*G the angle RCG of 73 degrees

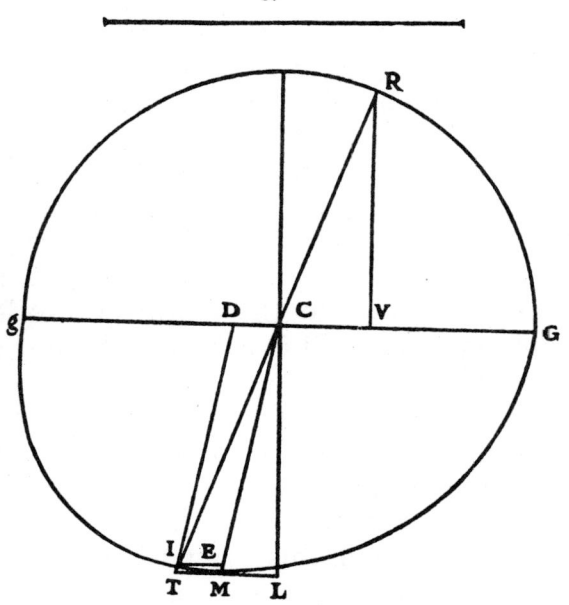

73 degrees 20 minutes, inclining to the same side as the crystal (of which ray mention has been made above); if one investigates, by the process above explained, the refraction CI, one will find that it makes exactly a straight line with RC, and that thus this ray is not deviated at all, conformably with experiment. This is proved as follows by calculation.

CG or CR being, as precedently, 98,779; CM being 100,000; and the angle RCV 73 degrees 20 minutes, CV will be 28,330. But because CI is the refraction of the ray RC, the proportion of CV to CD is 156,962 to 98,779, namely, that of N to CG; then CD is 17,828.

Now the rectangle gDC is to the square of DI as the square of CG is to the square of CM; hence DI or CE will be 98,353. But as CE is to EI, so will CM be to MT, which will then be 18,127. And being added to ML, which is 11,609 (namely the sine of the angle LCM, which is 6 degrees 40 minutes, taking CM 100,000 as radius) we get LT 27,936; and this is to LC 99,324 as CV to VR, that is to say, as 29,938, the tangent of the complement of the angle RCV, which is 73 degrees 20 minutes, is to the radius of the Tables. Whence it appears that RCIT is a straight line; which was to be proved.

35. Further it will be seen that the ray CI in emerging through the opposite surface of the crystal, ought to pass out quite straight, according to the following demonstration, which proves that the reciprocal relation of refraction obtains in this crystal the same as in other transparent bodies; that is to say, that if a ray RC in meeting the surface of the crystal CG is refracted as CI, the ray CI emerging through the opposite parallel surface of the
crystal

crystal, which I suppose to be IB, will have its refraction IA parallel to the ray RC.

Let the same things be supposed as before; that is to say, let CO, perpendicular to CR, represent a portion of a wave the continuation of which in the crystal is IK, so that the piece C will be continued on along the straight line CI, while O comes to K. Now if one takes a second period of time equal to the first, the piece K of the wave IK will, in this second period, have advanced along the straight line KB, equal and parallel to CI, because every piece of the wave CO, on arriving at the surface CK, ought to go on in the crystal the same as the piece C; and in this same time there will be formed in the air from the point I a partial spherical wave having a semi-diameter IA equal to KO, since KO has been traversed in an equal time. Similarly, if one considers some other point of the wave IK, such as *h*, it will go along *hm*, parallel to CI', to meet the surface IB, while the point K traverses K*l* equal to *hm*; and while this accomplishes the remainder *l*B, there will start from the point *m* a partial wave the semi-diameter of which, *mn*, will have the same ratio to *l*B as IA to KB.

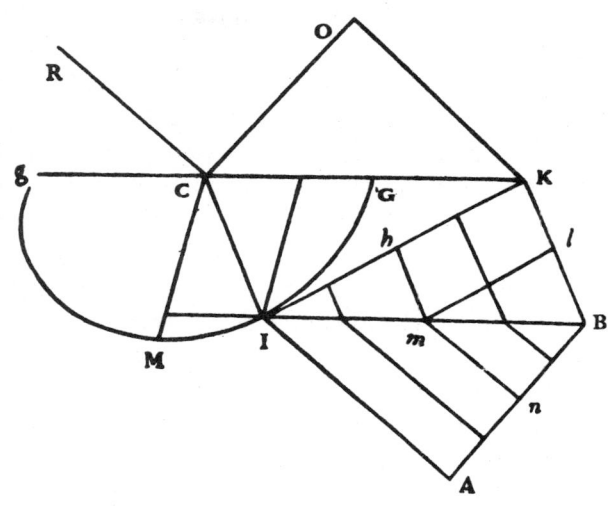

KB. Whence it is evident that this wave of semi-diameter *mn*, and the other of semi-diameter IA will have the same tangent BA. And similarly for all the partial spherical waves which will be formed outside the crystal by the impact of all the points of the wave IK against the surface of the Ether IB. It is then precisely the tangent BA which will be the continuation of the wave IK, outside the crystal, when the piece K has reached B. And in consequence IA, which is perpendicular to BA, will be the refraction of the ray CI on emerging from the crystal. Now it is clear that IA is parallel to the incident ray RC, since IB is equal to CK, and IA equal to KO, and the angles A and O are right angles.

It is seen then that, according to our hypothesis, the reciprocal relation of refraction holds good in this crystal as well as in ordinary transparent bodies; as is thus in fact found by observation.

36. I pass now to the consideration of other sections of the crystal, and of the refractions there produced, on which, as will be seen, some other very remarkable phenomena depend.

Let ABH be a parallelopiped of crystal, and let the top surface AEHF be a perfect rhombus, the obtuse angles of which are equally divided by the straight line EF, and the acute angles by the straight line AH perpendicular to FE.

The section which we have hitherto considered is that which passes through the lines EF, EB, and which at the same time cuts the plane AEHF at right angles. Refractions in this section have this in common with the refractions in ordinary media that the plane which is drawn through the incident ray and which also intersects the surface

surface of the crystal at right angles, is that in which the refracted ray also is found. But the refractions which appertain to every other section of this crystal have this strange property that the refracted ray always quits the plane of the incident ray perpendicular to the surface, and turns away towards the side of the slope of the crystal. For

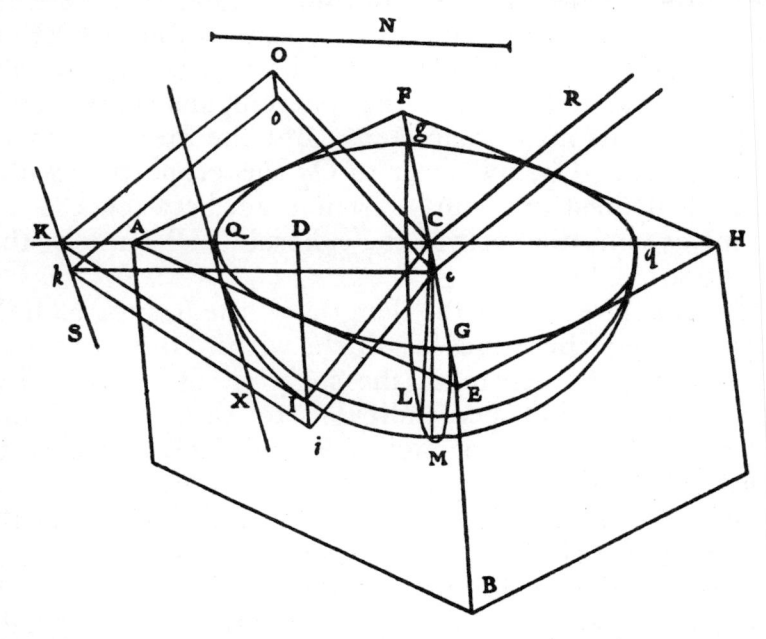

which fact we shall show the reason, in the first place, for the section through AH; and we shall show at the same time how one can determine the refraction, according to our hypothesis. Let there be, then, in the plane which passes through AH, and which is perpendicular to the plane AFHE, the incident ray RC; it is required to find its refraction in the crystal.

37. About

37. About the centre C, which I suppose to be in the intersection of AH and FE, let there be imagined a hemi-spheroid QG*q*gM, such as the light would form in spreading in the crystal, and let its section by the plane AEHF form the Ellipse QG*qg*, the major diameter of which Q*q*, which is in the line AH, will necessarily be one of the major diameters of the spheroid; because the axis of the spheroid being in the plane through FEB, to which QC is perpendicular, it follows that QC is also perpendicular to the axis of the spheroid, and consequently QC*q* one of its major diameters. But the minor diameter of this Ellipse, G*g*, will bear to Q*q* the proportion which has been defined previously, Article 27, between CG and the major semi-diameter of the spheroid, CP, namely, that of 98,779 to 105,032.

Let the line N be the length of the travel of light in air during the time in which, within the crystal, it makes, from the centre C, the spheroid QG*q*gM. Then having drawn CO perpendicular to the ray CR and situate in the plane through CR and AH, let there be adjusted, across the angle ACO, the straight line OK equal to N and perpendicular to CO, and let it meet the straight line AH at K. Supposing consequently that CL is perpendicular to the surface of the crystal AEHF, and that CM is the refraction of the ray which falls perpendicularly on this same surface, let there be drawn a plane through the line CM and through KCH, making in the spheroid the semi-ellipse QM*q*, which will be given, since the angle MCL is given of value 6 degrees 40 minutes. And it is certain, according to what has been explained above, Article 27, that a plane which would touch the spheroid at the point M, where I suppose the
straight

ON LIGHT. Chap. V

straight line CM to meet the surface, would be parallel to the plane QG*q*. If then through the point K one now draws KS parallel to G*g*, which will be parallel also to QX, the tangent to the Ellipse QG*q* at Q; and if one conceives a plane passing through KS and touching the spheroid, the point of contact will necessarily be in the Ellipse QM*q*, because this plane through KS, as well as the plane which touches the spheroid at the point M, are parallel to QX, the tangent of the spheroid : for this consequence will be demonstrated at the end of this Treatise. Let this point of contact be at I, then making KC, QC, DC proportionals, draw DI parallel to CM; also join CI. I say that CI will be the required refraction of the ray RC. This will be manifest if, in considering CO, which is perpendicular to the ray RC, as a portion of the wave of light, we can demonstrate that the continuation of its piece C will be found in the crystal at I, when O has arrived at K.

38. Now as in the Chapter on Reflexion, in demonstrating that the incident and reflected rays are always in the same plane perpendicular to the reflecting surface, we considered the breadth of the wave of light, so, similarly, we must here consider the breadth of the wave CO in the diameter G*g*. Taking then the breadth C*c* on the side toward the angle E, let the parallelogram CO*oc* be taken as a portion of a wave, and let us complete the parallelograms CK*kc*, CI*ic*, KI*ik*, OK*ko*. In the time then that the line O*o* arrives at the surface of the crystal at K*k*, all the points of the wave CO*oc* will have arrived at the rectangle K*c* along lines parallel to OK; and from the points of their incidences there will originate, beyond that, in the crystal partial hemi-spheroids, similar to the hemi-

hemi-spheroid Q*M*q, and similarly disposed. These hemi-spheroids will necessarily all touch the plane of the parallelogram KI*ik* at the same instant that O*o* has reached K*k*. Which is easy to comprehend, since, of these hemi-spheroids, all those which have their centres along the line CK, touch this plane in the line KI (for this is to be shown in the same way as we have demonstrated the refraction of the oblique ray in the principal section through EF) and all those which have their centres in the line C*c* will touch the same plane KI in the line I*i*; all these being similar to the hemi-spheroid Q*M*q. Since then the parallelogram K*i* is that which touches all these spheroids, this same parallelogram will be precisely the continuation of the wave CO*oc* in the crystal, when O*o* has arrived at K*k*, because it forms the termination of the movement and because of the quantity of movement which occurs more there than anywhere else: and thus it appears that the piece C of the wave CO*oc* has its continuation at I; that is to say, that the ray RC is refracted as CI.

From this it is to be noted that the proportion of the refraction for this section of the crystal is that of the line N to the semi-diameter CQ; by which one will easily find the refractions of all incident rays, in the same way as we have shown previously for the case of the section through FE; and the demonstration will be the same. But it appears that the said proportion of the refraction is less here than in the section through FEB; for it was there the same as the ratio of N to CG, that is to say, as 156,962 to 98,779, very nearly as 8 to 5; and here it is the ratio of N to CQ the major semi-diameter of the spheroid, that is to say, as 156,962 to 105,032, very nearly
as

as 3 to 2, but just a little less. Which still agrees perfectly with what one finds by observation.

39. For the rest, this diversity of proportion of refraction produces a very singular effect in this Crystal; which is that when it is placed upon a sheet of paper on which there are letters or anything else marked, if one views it from above with the two eyes situated in the plane of the section through EF, one sees the letters raised up by this irregular refraction more than when one puts one's eyes in the plane of section through AH : and the difference of these elevations appears by comparison with the other ordinary refraction of the crystal, the proportion of which is as 5 to 3, and which always raises the letters equally, and higher than the irregular refraction does. For one sees the letters and the paper on which they are written, as on two different stages at the same time; and in the first position of the eyes, namely, when they are in the plane through AH these two stages are four times more distant from one another than when the eyes are in the plane through EF.

We will show that this effect follows from the refractions; and it will enable us at the same time to ascertain the apparent place of a point of an object placed immediately under the crystal, according to the different situation of the eyes.

40. Let us see first by how much the irregular refraction of the plane through AH ought to lift the bottom of the crystal. Let the plane of this figure represent separately the section through Qq and CL, in which section there is also the ray RC, and let the semi-elliptic plane through Qq and CM be inclined to the former, as previously, by an angle of 6 degrees 40 minutes; and in this plane CI is then the refraction of the ray RC.

If now one considers the point I as at the bottom of the crystal, and that it is viewed by the rays ICR, I*cr*, refracted equally at the points C*c*, which should be equally distant from D, and that these rays meet the two eyes at R*r*; it is certain that the point I will appear raised to S where the straight lines RC, *rc*, meet; which point S is in DP, perpendicular to Q*q*. And if upon DP there is drawn the perpendicular IP, which will lie at the bottom of the crystal, the length SP will be the apparent elevation of the point I above the bottom.

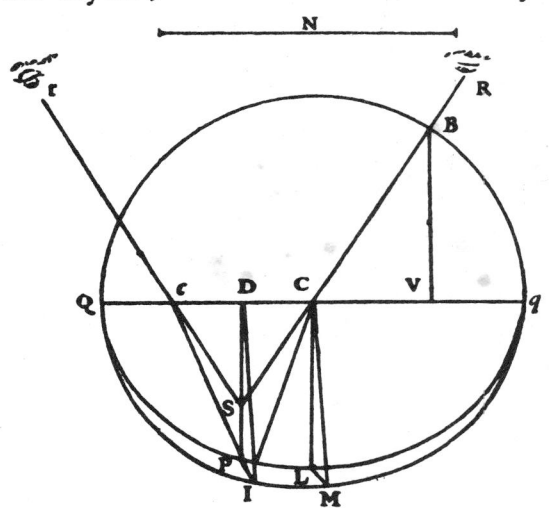

Let there be described on Q*q* a semicircle cutting the ray CR at B, from which BV is drawn perpendicular to Q*q*; and let the proportion of the refraction for this section be, as before, that of the line N to the semi-diameter CQ.

Then as N is to CQ so is VC to CD, as appears by the method of finding the refraction which we have shown above, Article 31; but as VC is to CD, so is VB to DS. Then as N is to CQ, so is VB to DS. Let ML be perpendicular to CL. And because I suppose the eyes R*r* to be distant about a foot or so from the crystal, and consequently the angle RS*r* very small, VB may be considered as equal to the semi-diameter CQ, and DP as equal to CL; then as N is to CQ

ON LIGHT. Chap. V

CQ so is CQ to DS. But N is valued at 156,962 parts, of which CM contains 100,000 and CQ 105,032. Then DS will have 70,283. But CL is 99,324, being the sine of the complement of the angle MCL which is 6 degrees 40 minutes; CM being supposed as radius. Then DP, considered as equal to CL, will be to DS as 99,324 to 70,283. And so the elevation of the point I by the refraction of this section is known.

41. Now let there be represented the other section through EF in the figure before the preceding one; and let CMg be the semi-ellipse, considered in Articles 27 and 28, which is made by cutting a spheroidal wave having centre C. Let the point I, taken in this ellipse, be imagined again at the bottom of the Crystal; and let it be viewed by the refracted rays ICR, Icr, which go to the two eyes; CR and cr being equally inclined to the surface of the crystal Gg. This being so, if one draws ID parallel to CM, which I suppose to be the refraction of the perpendicular ray incident at the point C, the distances DC, Dc, will be equal, as is easy to see by that which has been demonstrated in Article 28. Now it is certain that the point I should appear at S where the straight lines RC, rc, meet when prolonged; and that this point will fall in the line

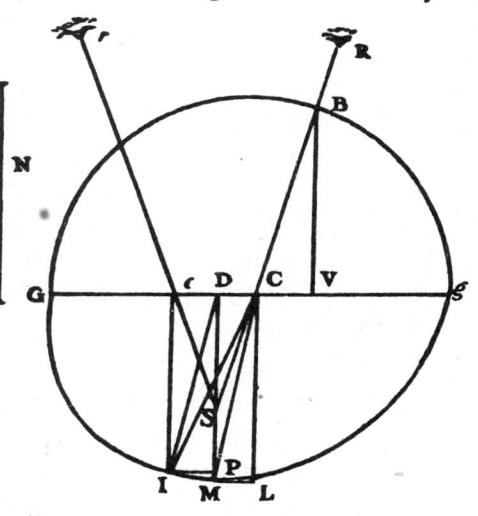

84 TREATISE

line DP perpendicular to Gg. If one draws IP perpendicular to this DP, it will be the distance PS which will mark the apparent elevation of the point I. Let there be described on Gg a semicircle cutting CR at B, from which let BV be drawn perpendicular to Gg; and let N to GC be the proportion of the refraction in this section, as in Article 28. Since then CI is the refraction of the radius BC, and DI is parallel to CM, VC must be to CD as N to GC, according to what has been demonstrated in Article 31. But as VC is to CD so is BV to DS. Let ML be drawn perpendicular to CL. And because I consider, again, the eyes to be distant above the crystal, BV is deemed equal to the semi-diameter CG; and hence DS will be a third proportional to the lines N and CG: also DP will be deemed equal to CL. Now CG consisting of 98,778 parts, of which CM contains 100,000, N is taken as 156,962. Then DS will be 62,163. But CL is also determined, and contains 99,324 parts, as has been said in Articles 34 and 40. Then the ratio of PD to DS will be as 99,324 to 62,163. And thus one knows the elevation of the point at the bottom I by the refraction of this section; and it appears that this elevation is greater than that by the refraction of the preceding section, since the ratio of PD to DS was there as 99,324 to 70,283.

But by the regular refraction of the crystal, of which we have above said that the proportion is 5 to 3, the elevation of the point I, or P, from the bottom, will be $\frac{2}{5}$ of the height DP; as appears by this figure, where the point P being viewed by the rays PCR, P*cr*, refracted equally

at

at the surface C*c*, this point must needs appear to be at S, in the perpendicular PD where the lines RC, *rc*, meet when prolonged: and one knows that the line PC is to CS as 5 to 3, since they are to one another as the sine of the angle CSP or DSC is to the sine of the angle SPC. And because the ratio of PD to DS is deemed the same as that of PC to CS, the two eyes R*r* being supposed very far above the crystal, the elevation PS will thus be $\frac{2}{5}$ of PD.

42. If one takes a straight line AB for the thickness of the crystal, its point B being at the bottom, and if one divides it at the points C, D, E, according to the proportions of the elevations found, making AE $\frac{3}{5}$ of AB, AB to AC as 99,324 to 70,283, and AB to AD as 99,324 to 62,163, these points will divide AB as in this figure. And it will be found that this agrees perfectly with experiment; that is to say by placing the eyes above in the plane which cuts the crystal according to the shorter diameter of the rhombus, the regular refraction will lift up the letters to E; and one will see the bottom, and the letters over which it is placed, lifted up to D by the irregular refraction. But by placing the eyes above in the plane which cuts the crystal according to the longer diameter of the rhombus, the regular refraction will lift the letters to E as before; but the irregular refraction will make them, at the same time, appear lifted up only to C; and in such a way that the interval CE will be quadruple the interval ED, which one previously saw.

43. I have only to make the remark here that in both the positions of the eyes the images caused by the irregular refraction do not appear directly below those which proceed

ceed from the regular refraction, but they are separated from them by being more distant from the equilateral solid angle of the Crystal. That follows, indeed, from all that has been hitherto demonstrated about the irregular refraction; and it is particularly shown by these last demonstrations, from which one sees that the point I appears by irregular refraction at S in the perpendicular line DP, in which line also the image of the point P ought to appear by regular refraction, but not the image of the point I, which will be almost directly above the same point, and higher than S.

But as to the apparent elevation of the point I in other positions of the eyes above the crystal, besides the two positions which we have just examined, the image of that point by the irregular refraction will always appear between the two heights of D and C, passing from one to the other as one turns one's self around about the immovable crystal, while looking down from above. And all this is still found conformable to our hypothesis, as any one can assure himself after I shall have shown here the way of finding the irregular refractions which appear in all other sections of the crystal, besides the two which we have considered. Let us suppose one of the faces of the crystal, in which let there be the Ellipse HDE, the centre C of which is also the centre of the spheroid HME in which the light spreads, and of which the said Ellipse is the section. And let the incident ray be RC, the refraction of which it is required to find.

Let there be taken a plane passing through the ray RC and which is perpendicular to the plane of the ellipse HDE, cutting it along the straight line BCK; and having in the same plane through RC made CO perpendicular to CR,
let

let OK be adjusted across the angle OCK, so as to be perpendicular to OC and equal to the line N, which I suppose to measure the travel of the light in air during the time that it spreads in the crystal through the spheroid HDEM. Then in the plane of the Ellipse HDE let KT be drawn, through the point K, perpendicular to BCK. Now if one conceives a plane drawn through the straight line KT and touching the spheroid HME at I, the straight line CI will be the refraction of the ray RC, as is easy to deduce from that which has been demonstrated in Article 36.

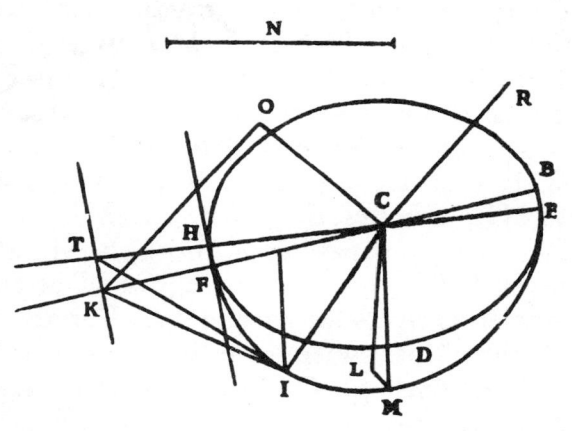

But it must be shown how one can determine the point of contact I. Let there be drawn parallel to the line KT a line HF which touches the Ellipse HDE, and let this point of contact be at H. And having drawn a straight line along CH to meet KT at T, let there be imagined a plane passing through the same CH and through CM (which I suppose to be the refraction of the perpendicular ray), which makes in the spheroid the elliptical section HME. It is certain that the plane which will pass through the straight line KT, and which will touch the spheroid, will touch it at a point in the Ellipse HME, according to the Lemma which will be demonstrated at the end of the Chapter.

Chapter. Now this point is necessarily the point I which is sought, since the plane drawn through TK can touch the spheroid at one point only. And this point I is easy to determine, since it is needful only to draw from the point T, which is in the plane of this Ellipse, the tangent TI, in the way shown previously. For the Ellipse HME is given, and its conjugate semi-diameters are CH and CM; because a straight line drawn through M, parallel to HE, touches the Ellipse HME, as follows from the fact that a plane taken through M, and parallel to the plane HDE, touches the spheroid at that point M, as is seen from Articles 27 and 23. For the rest, the position of this ellipse, with respect to the plane through the ray RC and through CK, is also given; from which it will be easy to find the position of CI, the refraction corresponding to the ray RC.

Now it must be noted that the same ellipse HME serves to find the refractions of any other ray which may be in the plane through RC and CK. Because every plane, parallel to the straight line HF, or TK, which will touch the spheroid, will touch it in this ellipse, according to the Lemma quoted a little before.

I have investigated thus, in minute detail, the properties of the irregular refraction of this Crystal, in order to see whether each phenomenon that is deduced from our hypothesis accords with that which is observed in fact. And this being so it affords no slight proof of the truth of our suppositions and principles. But what I am going to add here confirms them again marvellously. It is this: that there are different sections of this Crystal, the surfaces of which, thereby produced, give rise to refractions precisely such as they ought to be, and as I had foreseen them, according to the preceding Theory.

In

ON LIGHT. Chap. V

In order to explain what these sections are, let ABKF be the principal section through the axis of the crystal ACK, in which there will also be the axis SS of a spheroidal wave of light spreading in the crystal from the centre C; and the straight line which cuts SS through the middle and at right angles, namely PP, will be one of the major diameters.

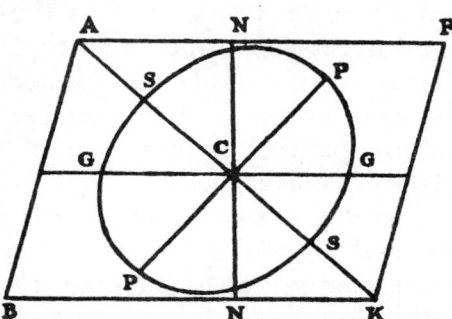

Now as in the natural section of the crystal, made by a plane parallel to two opposite faces, which plane is here represented by the line GG, the refraction of the surfaces which are produced by it will be governed by the hemi-spheroids GNG, according to what has been explained in the preceding Theory. Similarly, cutting the Crystal through NN, by a plane perpendicular to the parallelogram ABKF, the refraction of the surfaces will be governed by the hemi-spheroids NGN. And if one cuts it through PP, perpendicularly to the said parallelogram, the refraction of the surfaces ought to be governed by the hemi-spheroids PSP, and so for others. But I saw that if the plane NN was almost perpendicular to the plane GG, making the angle NCG, which is on the side A, an angle of 90 degrees 40 minutes, the hemi-spheroids NGN would become similar to the hemi-spheroids GNG, since the planes NN and GG were equally inclined by an angle of 45 degrees 20 minutes to the axis SS. In consequence it must needs be, if our theory is true, that the surfaces which the section through

NN produces should effect the same refractions as the surfaces of the section through GG. And not only the surfaces of the section NN but all other sections produced by planes which might be inclined to the axis at an angle equal to 45 degrees 20 minutes. So that there are an infinitude of planes which ought to produce precisely the same refractions as the natural surfaces of the crystal, or as the section parallel to any one of those surfaces which are made by cleavage.

I saw also that by cutting it by a plane taken through PP, and perpendicular to the axis SS, the refraction of the surfaces ought to be such that the perpendicular ray should suffer thereby no deviation; and that for oblique rays there would always be an irregular refraction, differing from the regular, and by which objects placed beneath the crystal would be less elevated than by that other refraction.

That, similarly, by cutting the crystal by any plane through the axis SS, such as the plane of the figure is, the perpendicular ray ought to suffer no refraction; and that for oblique rays there were different measures for the irregular refraction according to the situation of the plane in which the incident ray was.

Now these things were found in fact so; and, after that, I could not doubt that a similar success could be met with everywhere. Whence I concluded that one might form from this crystal solids similar to those which are its natural forms, which should produce, at all their surfaces, the same regular and irregular refractions as the natural surfaces, and which nevertheless would cleave in quite other ways, and not in directions parallel to any of their faces. That out of it one would be able to fashion pyramids, having their base square, pentagonal, hexagonal, or with as many sides as as one desired, all the surfaces of which should have the same refractions as the natural surfaces of the crystal, except the base, which will not refract the perpendicular ray. These surfaces will each make an angle of 45 degrees 20 minutes with the axis of the crystal, and the base will be the section perpendicular to the axis.

That, finally, one could also fashion out of it triangular prisms, or prisms with as many sides as one would, of which neither the sides nor the bases would refract the perpendicular ray, although they would yet all cause double refraction for oblique rays. The cube is included amongst these prisms, the bases of which are sections perpendicular to the axis of the crystal, and the sides are sections parallel to the same axis.

From all this it further appears that it is not at all in the disposition of the layers of which this crystal seems to be composed, and according to which it splits in three different senses, that the cause resides of its irregular refraction; and that it would be in vain to wish to seek it there.

But in order that any one who has some of this stone may be able to find, by his own experience, the truth of what I have just advanced, I will state here the process of which I have made use to cut it, and to polish it. Cutting is easy by the slicing wheels of lapidaries, or in the way in which marble is sawn: but polishing is very difficult, and by employing the ordinary means one more often depolishes the surfaces than makes them lucent.

After many trials, I have at last found that for this service no plate of metal must be used, but a piece of mirror glass made matt and depolished. Upon this, with fine sand and water, one smoothes the crystal little by little, in the same way

way as spectacle glasses, and polishes it simply by continuing the work, but ever reducing the material. I have not, however, been able to give it perfect clarity and transparency; but the evenness which the surfaces acquire enables one to observe in them the effects of refraction better than in those made by cleaving the stone, which always have some inequality.

Even when the surface is only moderately smoothed, if one rubs it over with a little oil or white of egg, it becomes quite transparent, so that the refraction is discerned in it quite distinctly. And this aid is specially necessary when it is wished to polish the natural surfaces to remove the inequalities; because one cannot render them lucent equally with the surfaces of other sections, which take a polish so much the better the less nearly they approximate to these natural planes.

Before finishing the treatise on this Crystal, I will add one more marvellous phenomenon which I discovered after having written all the foregoing. For though I have not been able till now to find its cause, I do not for that reason wish to desist from describing it, in order to give opportunity to others to investigate it. It seems that it will be necessary to make still further suppositions besides those which I have made; but these will not for all that cease to keep their probability after having been confirmed by so many tests.

The phenomenon is, that by taking two pieces of this crystal and applying them one over the other, or rather holding them with a space between the two, if all the sides of one are parallel to those of the other, then a ray of light, such as AB, is divided into two in the first piece, namely into BD and BC, following the two refractions,
regular

ON LIGHT. Chap. V

regular and irregular. On penetrating thence into the other piece each ray will pass there without further divid-

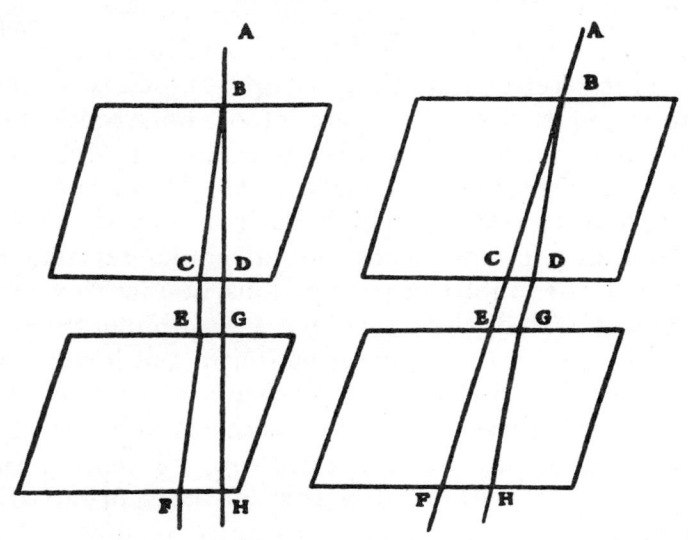

ing itself in two; but that one which underwent the regular refraction, as here DG, will undergo again only a regular refraction at GH; and the other, CE, an irregular refraction at EF. And the same thing occurs not only in this disposition, but also in all those cases in which the principal section of each of the pieces is situated in one and the same plane, without it being needful for the two neighbouring surfaces to be parallel. Now it is marvellous why the rays CE and DG, incident from the air on the lower crystal, do not divide themselves the same as the first ray AB. One would say that it must be that the ray DG in passing through the upper piece has lost something which is necessary to move the matter which serves for the irregular refraction; and that likewise CE has lost that which

was

was necessary to move the matter which serves for regular refraction: but there is yet another thing which upsets this reasoning. It is that when one disposes the two crystals in such a way that the planes which constitute the principal sections intersect one another at right angles, whether the neighbouring surfaces are parallel or not, then the ray which has come by the regular refraction, as DG, undergoes only an irregular refraction in the lower piece; and on the contrary the ray which has come by the irregular refraction, as CE, undergoes only a regular refraction.

But in all the infinite other positions, besides those which I have just stated, the rays DG, CE, divide themselves anew each one into two, by refraction in the lower crystal, so that from the single ray AB there are four, sometimes of equal brightness, sometimes some much less bright than others, according to the varying agreement in the positions of the crystals: but they do not appear to have all together more light than the single ray AB.

When one considers here how, while the rays CE, DG, remain the same, it depends on the position that one gives to the lower piece, whether it divides them both in two, or whether it does not divide them, and yet how the ray AB above is always divided, it seems that one is obliged to conclude that the waves of light, after having passed through the first crystal, acquire a certain form or disposition in virtue of which, when meeting the texture of the second crystal, in certain positions, they can move the two different kinds of matter which serve for the two species of refraction; and when meeting the second crystal in another position are able to move only one of these kinds of matter. But to tell how this occurs, I have hitherto found nothing which satisfies me.

Leaving

Leaving then to others this research, I pass to what I have to say touching the cause of the extraordinary figure of this crystal, and why it cleaves easily in three different senses, parallel to any one of its surfaces.

There are many bodies, vegetable, mineral, and congealed salts, which are formed with certain regular angles and figures. Thus among flowers there are many which have their leaves disposed in ordered polygons, to the number of 3, 4, 5, or 6 sides, but not more. This well deserves to be investigated, both as to the polygonal figure, and as to why it does not exceed the number 6.

Rock Crystal grows ordinarily in hexagonal bars, and diamonds are found which occur with a square point and polished surfaces. There is a species of small flat stones, piled up directly upon one another, which are all of pentagonal figure with rounded angles, and the sides a little folded inwards. The grains of gray salt which are formed from sea water affect the figure, or at least the angle, of the cube; and in the congelations of other salts, and in that of sugar, there are found other solid angles with perfectly flat faces. Small snowflakes almost always fall in little stars with 6 points, and sometimes in hexagons with straight sides. And I have often observed, in water which is beginning to freeze, a kind of flat and thin foliage of ice, the middle ray of which throws out branches inclined at an angle of 60 degrees. All these things are worthy of being carefully investigated to ascertain how and by what artifice nature there operates. But it is not now my intention to treat fully of this matter. It seems that in general the regularity which occurs in these productions comes from the arrangement of the small invisible equal particles of which they are composed. And, coming to our Iceland Crystal, I say that

that if there were a pyramid such as ABCD, composed of small rounded corpuscles, not spherical but flattened spheroids, such as would be made by the rotation of the ellipse GH around its lesser diameter EF (of which the ratio to the greater diameter is very nearly that of 1 to the square root of 8)—I say that then the solid angle of the point D would be equal to the obtuse and equilateral angle of this Crystal. I say, further, that if these corpuscles were lightly stuck together, on breaking this pyramid it would break along faces parallel to those that make its point: and by this means, as it is easy to see, it would produce prisms similar to those of the same crystal as this other figure represents. The reason is that when broken in this fashion a whole layer separates easily from its neighbouring layer since each spheroid has to be detached only from the three spheroids of the next layer; of which three there is but one which touches it on its flattened surface, and the other two at the edges. And the reason why the surfaces separate sharp and polished is that if any spheroid of the neighbouring surface would come out by attaching itself to the surface which is being separated, it would be needful for it to detach itself from six other spheroids which hold it locked, and four of which press it by these flattened surfaces. Since then not only the angles of our crystal but also the manner in which it splits agree precisely with what is observed in the assemblage composed of such spheroids, there is great reason to believe that the particles are shaped and ranged in the same way.

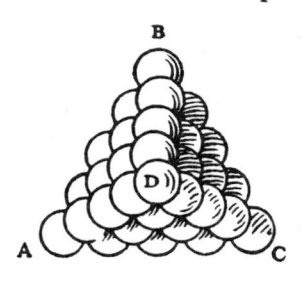

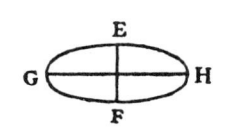

There

There is even probability enough that the prisms of this crystal are produced by the breaking up of pyramids, since Mr. Bartholinus relates that he occasionally found some pieces of triangularly pyramidal figure. But when a mass is composed interiorly only of these little spheroids thus piled up, whatever form it may have exteriorly, it is certain, by the same reasoning which I have just explained, that if broken it would produce similar prisms. It remains to be seen whether there are other reasons which confirm our conjecture, and whether there are none which are repugnant to it.

It may be objected that this crystal, being so composed, might be capable of cleavage in yet two more fashions; one of which would be along planes parallel to the base of the pyramid, that is to say to the triangle ABC; the other would be parallel to a plane the trace of which is marked by the lines GH, HK, KL. To which I say that both the one and the other, though practicable, are more difficult than those which were parallel to any one of the three planes of the pyramid; and that therefore, when striking on the crystal in order to break it, it ought always to split rather along these three planes than along the two others. When one has a number of spheroids of the form above described, and ranges them in a pyramid, one sees why the two methods of division are more difficult. For in the case of that division which would be parallel to the base,

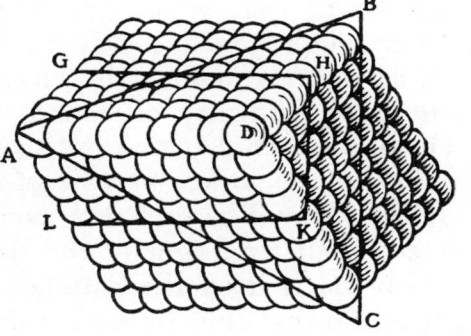

each

each spheroid would be obliged to detach itself from three others which it touches upon their flattened surfaces, which hold more strongly than the contacts at the edges. And besides that, this division will not occur along entire layers, because each of the spheroids of a layer is scarcely held at all by the 6 of the same layer that surround it, since they only touch it at the edges; so that it adheres readily to the neighbouring layer, and the others to it, for the same reason; and this causes uneven surfaces. Also one sees by experiment that when grinding down the crystal on a rather rough stone, directly on the equilateral solid angle, one verily finds much facility in reducing it in this direction, but much difficulty afterwards in polishing the surface which has been flattened in this manner.

As for the other method of division along the plane GHKL, it will be seen that each spheroid would have to detach itself from four of the neighbouring layer, two of which touch it on the flattened surfaces, and two at the edges. So that this division is likewise more difficult than that which is made parallel to one of the surfaces of the crystal; where, as we have said, each spheroid is detached from only three of the neighbouring layer: of which three there is one only which touches it on the flattened surface, and the other two at the edges only.

However, that which has made me know that in the crystal there are layers in this last fashion, is that in a piece weighing half a pound which I possess, one sees that it is split along its length, as is the above-mentioned prism by the plane GHKL; as appears by colours of the Iris extending throughout this whole plane although the two pieces still hold together. All this proves then that the composition of the crystal is such as we have stated. To which

which I again add this experiment; that if one passes a knife scraping along any one of the natural surfaces, and downwards as it were from the equilateral obtuse angle, that is to say from the apex of the pyramid, one finds it quite hard; but by scraping in the opposite sense an incision is easily made. This follows manifestly from the situation of the small spheroids; over which, in the first manner, the knife glides; but in the other manner it seizes them from beneath almost as if they were the scales of a fish.

I will not undertake to say anything touching the way in which so many corpuscles all equal and similar are generated, nor how they are set in such beautiful order; whether they are formed first and then assembled, or whether they arrange themselves thus in coming into being and as fast as they are produced, which seems to me more probable. To develop truths so recondite there would be needed a knowledge of nature much greater than that which we have. I will add only that these little spheroids could well contribute to form the spheroids of the waves of light, here above supposed, these as well as those being similarly situated, and with their axes parallel.

Calculations which have been supposed in this Chapter.

Mr. Bartholinus, in his treatise of this Crystal, puts at 101 degrees the obtuse angles of the faces, which I have stated to be 101 degrees 52 minutes. He states that he measured these angles directly on the crystal, which is difficult to do with ultimate exactitude, because the edges such as CA, CB, in this figure, are generally worn, and not quite straight. For more certainty, therefore, I preferred to measure actually the obtuse angle by which the faces CBDA,

CBDA, CBVF, are inclined to one another, namely the angle OCN formed by drawing CN perpendicular to FV, and CO perpendicular to DA. This angle OCN I found to be 105 degrees; and its supplement CNP, to be 75 degrees, as it should be.

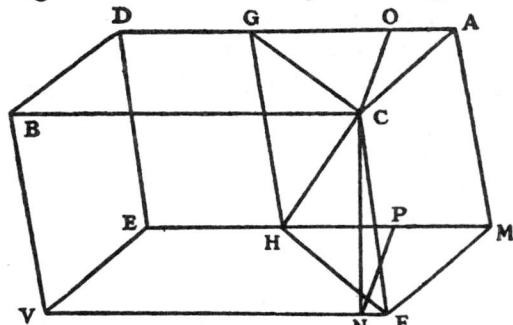

To find from this the obtuse angle BCA, I imagined a sphere having its centre at C, and on its surface a spherical triangle, formed by the intersection of three planes which enclose the solid angle C. In this equilateral triangle, which is ABF in this other figure, I see that each of the angles should be 105 degrees, namely equal to the angle OCN; and that each of the sides should be of as many degrees as the angle ACB, or ACF, or BCF. Having then drawn the arc FQ perpendicular to the side AB, which it divides equally at Q, the triangle FQA has a right angle at Q, the angle A 105 degrees, and F half as much, namely 52 degrees 30 minutes; whence the hypotenuse AF is found to be 101 degrees 52 minutes. And this arc AF is the measure of the angle ACF in the figure of the crystal.

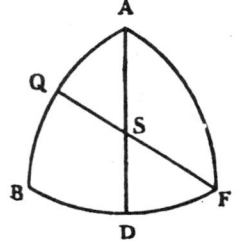

In the same figure, if the plane CGHF cuts the crystal so that it divides the obtuse angles ACB, MHV, in the middle, it is stated, in Article 10, that the angle CFH is 70 degrees 57 minutes. This again is easily shown in the same

same spherical triangle ABF, in which it appears that the arc FQ is as many degrees as the angle GCF in the crystal, the supplement of which is the angle CFH. Now the arc FQ is found to be 109 degrees 3 minutes. Then its supplement, 70 degrees 57 minutes, is the angle CFH.

It was stated, in Article 26, that the straight line CS, which in the preceding figure is CH, being the axis of the crystal, that is to say being equally inclined to the three sides CA, CB, CF, the angle GCH is 45 degrees 20 minutes. This is also easily calculated by the same spherical triangle. For by drawing the other arc AD which cuts BF equally, and intersects FQ at S, this point will be the centre of the triangle. And it is easy to see that the arc SQ is the measure of the angle GCH in the figure which represents the crystal. Now in the triangle QAS, which is right-angled, one knows also the angle A, which is 52 degrees 30 minutes, and the side AQ 50 degrees 56 minutes; whence the side SQ is found to be 45 degrees 20 minutes.

In Article 27 it was required to show that PMS being an ellipse the centre of which is C, and which touches the straight line MD at M so that the angle MCL which CM makes with CL, perpendicular on DM, is 6 degrees 40 minutes, and its semi-minor axis CS making with CG (which is parallel to MD) an angle GCS of 45 degrees 20 minutes, it was required to show, I say, that, CM being 100,000 parts, PC the semi-major diameter of this ellipse is 105,032 parts, and CS, the semi-minor diameter, 93,410.

Let CP and CS be prolonged and meet the tangent DM at D and Z; and from the point of contact M let MN and MO be drawn as perpendiculars to CP and CS. Now because the angles SCP, GCL, are right angles, the
angle

angle PCL will be equal to GCS which was 45 degrees 20 minutes. And deducting the angle LCM, which is 6 degrees 40 minutes, from LCP, which is 45 degrees 20 minutes, there remains MCP, 38 degrees 40 minutes. Considering then CM as a radius of 100,000 parts, MN, the sine of 38 degrees 40 minutes, will be 62,479. And in the right-angled triangle MND, MN will be to ND as the radius of the Tables is to the tangent of 45 degrees 20 minutes (because the angle NMD is equal to DCL, or GCS); that is to say as 100,000 to 101,170: whence results ND 63,210. But NC is 78,079 of the same parts, CM being 100,000, because NC is the sine of the complement of the angle MCP, which was 38 degrees 40 minutes. Then the whole line DC is 141,289; and CP, which is a mean proportional between DC and CN, since MD touches the Ellipse, will be 105,032.

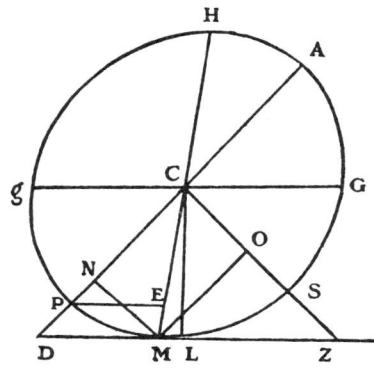

Similarly, because the angle OMZ is equal to CDZ, or LCZ, which is 44 degrees 40 minutes, being the complement of GCS, it follows that, as the radius of the Tables is to the tangent of 44 degrees 40 minutes, so will OM 78,079 be to OZ 77,176. But OC is 62,479 of these same parts of which CM is 100,000, because it is equal to MN, the sine of the angle MCP, which is 38 degrees 40 minutes. Then the whole line CZ is 139,655; and CS, which is a mean proportional between CZ and CO will be 93,410.

At

ON LIGHT. Chap. V

At the same place it was stated that GC was found to be 98,779 parts. To prove this, let PE be drawn in the same figure parallel to DM, and meeting CM at E. In the right-angled triangle CLD the side CL is 99,324 (CM being 100,000), because CL is the sine of the complement of the angle LCM, which is 6 degrees 40 minutes. And since the angle LCD is 45 degrees 20 minutes, being equal to GCS, the side LD is found to be 100,486 : whence deducting ML 11,609 there will remain MD 88,877. Now as CD (which was 141,289) is to DM 88,877, so will CP 105,032 be to PE 66,070. But as the rectangle MEH (or rather the difference of the squares on CM and CE) is to the square on MC, so is the square on PE to the square on Cg; then also as the difference of the squares on DC and CP to the square on CD, so also is the square on PE to the square on gC. But DP, CP, and PE are known; hence also one knows GC, which is 98,779.

Lemma which has been supposed.

If a spheroid is touched by a straight line, and also by two or more planes which are parallel to this line, though not parallel to one another, all the points of contact of the line, as well as of the planes, will be in one and the same ellipse made by a plane which passes through the centre of the spheroid.

Let LED be the spheroid touched by the line BM at the point B, and also by the planes parallel to this line at the points O and A. It is required to demonstrate that the points B, O, and A are in one and the same Ellipse made in the spheroid by a plane which passes through its centre.

Through

Through the line BM, and through the points O and A, let there be drawn planes parallel to one another, which, in cutting the spheroid make the ellipses LBD, POP, QAQ; which will all be similar and similarly disposed, and will have their centres K, N, R, in one and the same diameter of the spheroid, which will also be the diameter of the ellipse made by the section of the plane that passes through the centre of the spheroid, and which cuts the planes of the three said Ellipses at right angles: for all this is manifest by proposition 15 of the book of Conoids and Spheroids of Archimedes. Further, the two latter planes, which are drawn through the points O and A, will also, by cutting the planes which touch the spheroid in these same points, generate straight lines, as OH and AS, which will, as is easy to see, be parallel to BM; and all three, BM, OH, AS, will touch the Ellipses LBD, POP, QAQ in these points, B, O, A; since they are in the planes of these ellipses, and at the same time in the planes which touch the spheroid. If now from these points B, O, A, there are drawn the straight lines BK, ON, AR, through the centres of the same ellipses, and if through these centres there are drawn also the diameters LD, PP, QQ, parallel to the tangents BM, OH, AS; these will be conjugate to the aforesaid BK, ON, AR. And because the three ellipses are similar and similarly disposed,

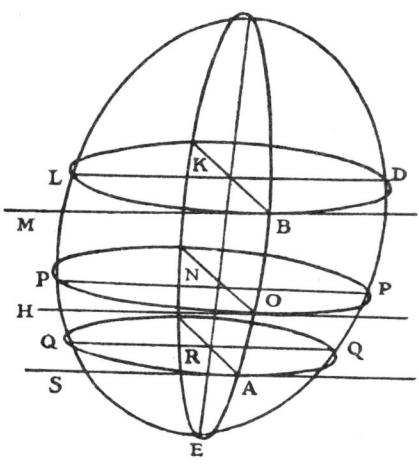

disposed, and have their diameters LD, PP, QQ parallel, it is certain that their conjugate diameters BK, ON, AR, will also be parallel. And the centres K, N, R being, as has been stated, in one and the same diameter of the spheroid, these parallels BK, ON, AR will necessarily be in one and the same plane, which passes through this diameter of the spheroid, and, in consequence, the points R, O, A are in one and the same ellipse made by the intersection of this plane. Which was to be proved. And it is manifest that the demonstration would be the same if, besides the points O, A, there had been others in which the spheroid had been touched by planes parallel to the straight line BM.

CHAPTER VI

On the Figures of the Transparent Bodies

Which serve for Refraction and for Reflexion.

AFTER having explained how the properties of reflexion and refraction follow from what we have supposed concerning the nature of light, and of opaque bodies, and of transparent media, I will here set forth a very easy and natural way of deducing, from the same principles, the true figures which serve, either by reflexion or by refraction, to collect or disperse the rays of light, as may be desired. For though I do not see yet that there are means of making use of these figures, so far as relates to Refraction, not only because of the difficulty of shaping the glasses of Telescopes with the requisite

quisite exactitude according to these figures, but also because there exists in refraction itself a property which hinders the perfect concurrence of the rays, as Mr. Newton has very well proved by experiment, I will yet not desist from relating the invention, since it offers itself, so to speak, of itself, and because it further confirms our Theory of refraction, by the agreement which here is found between the refracted ray and the reflected ray. Besides, it may occur that some one in the future will discover in it utilities which at present are not seen.

To proceed then to these figures, let us suppose first that it is desired to find a surface CDE which shall reassemble at a point B rays coming from another point A; and that the summit of the surface shall be the given point D in the straight line AB. I say that, whether by

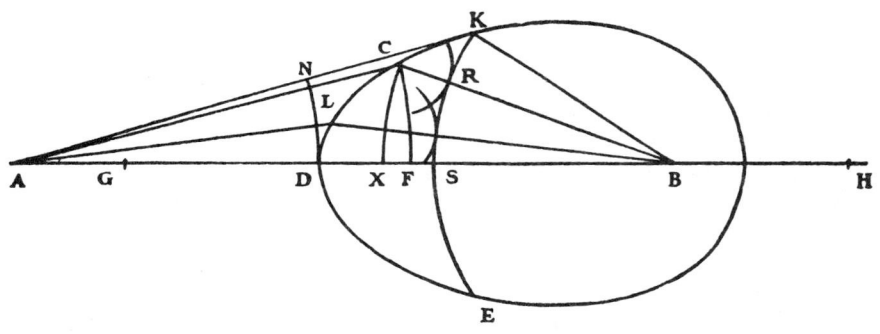

reflexion or by refraction, it is only necessary to make this surface such that the path of the light from the point A to all points of the curved line CDE, and from these to the point of concurrence (as here the path along the straight lines AC, CB, along AL, LB, and along AD, DB), shall be everywhere traversed in equal times: by which principle the finding of these curves becomes very easy.

So

ON LIGHT. Chap. VI

So far as relates to the reflecting surface, since the sum of the lines AC, CB ought to be equal to that of AD, DB, it appears that DCE ought to be an ellipse; and for refraction, the ratio of the velocities of waves of light in the media A and B being supposed to be known, for example that of 3 to 2 (which is the same, as we have shown, as the ratio of the Sines in the refraction), it is only necessary to make DH equal to $\frac{3}{2}$ of DB; and having after that described from the centre A some arc FC, cutting DB at F, then describe another from centre B with its semi-diameter BX equal to $\frac{2}{3}$ of FH; and the point of intersection of the two arcs will be one of the points required, through which the curve should pass. For this point, having been found in this fashion, it is easy forthwith to demonstrate that the time along AC, CB, will be equal to the time along AD, DB.

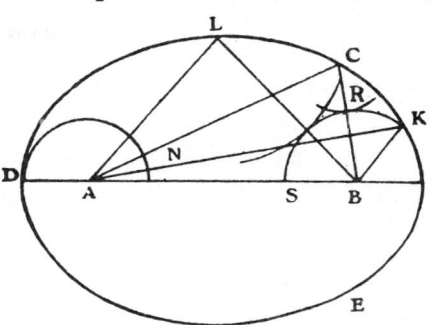

For assuming that the line AD represents the time which the light takes to traverse this same distance AD in air, it is evident that DH, equal to $\frac{3}{2}$ of DB, will represent the time of the light along DB in the medium, because it needs here more time in proportion as its speed is slower. Therefore the whole line AH will represent the time along AD, DB. Similarly the line AC or AF will represent the time along AC; and FH being by construction equal to $\frac{3}{2}$ of CB, it will represent the time along CB in the medium; and in consequence the whole line AH will represent also the time along AC, CB. Whence it appears that the time

time along AC, CB, is equal to the time along AD, DB. And similarly it can be shown if L and K are other points in the curve CDE, that the times along AL, LB, and along AK, KB, are always represented by the line AH, and therefore equal to the said time along AD, DB.

In order to show further that the surfaces, which these curves will generate by revolution, will direct all the rays which reach them from the point A in such wise that they tend towards B, let there be supposed a point K in the curve, farther from D than C is, but such that the straight line AK falls from outside upon the curve which serves for the refraction; and from the centre B let the arc KS be described, cutting BD at S, and the straight line CB at R; and from the centre A describe the arc DN meeting AK at N.

Since the sums of the times along AK, KB, and along AC, CB are equal, if from the former sum one deducts the time along KB, and if from the other one deducts the time along RB, there will remain the time along AK as equal to the time along the two parts AC, CR. Consequently in the time that the light has come along AK it will also have come along AC and will in addition have made, in the medium from the centre C, a partial spherical wave, having a semi-diameter equal to CR. And this wave will necessarily touch the circumference KS at R, since CB cuts this circumference at right angles. Similarly, having taken any other point L in the curve, one can show that in the same time as the light passes along AL it will also have come along AL and in addition will have made a partial wave, from the centre L, which will touch the same circumference KS. And so with all other points of the curve CDE. Then at the moment that the light reaches K the arc KRS will be the termination
of

of the movement, which has spread from A through DCK. And thus this same arc will constitute in the medium the propagation of the wave emanating from A; which wave may be represented by the arc DN, or by any other nearer the centre A. But all the pieces of the arc KRS are propagated successively along straight lines which are perpendicular to them, that is to say, which tend to the centre B (for that can be demonstrated in the same way as we have proved above that the pieces of spherical waves are propagated along the straight lines coming from their centre), and these progressions of the pieces of the waves constitute the rays themselves of light. It appears then that all these rays tend here towards the point B.

One might also determine the point C, and all the others, in this curve which serves for the refraction, by dividing DA at G in such a way that DG is $\frac{2}{3}$ of DA, and describing from the centre B any arc CX which cuts BD at X, and another from the centre A with its semi-diameter AF equal to $\frac{3}{2}$ of GX; or rather, having described, as before, the arc CX, it is only necessary to make DF equal to $\frac{3}{2}$ of DX, and from the centre A to strike the arc FC; for these two constructions, as may be easily known, come back to the first one which was shown before. And it is manifest by the last method that this curve is the same that Mr. Des Cartes has given in his Geometry, and which he calls the first of his Ovals.

It is only a part of this oval which serves for the refraction, namely, the part DK, ending at K, if AK is the tangent. As to the other part, Des Cartes has remarked that it could serve for reflexions, if there were some material of a mirror of such a nature that by its means

means the force of the rays (or, as we should say, the velocity of the light, which he could not say, since he held that the movement of light was instantaneous) could be augmented in the proportion of 3 to 2. But we have shown that in our way of explaining reflexion, such

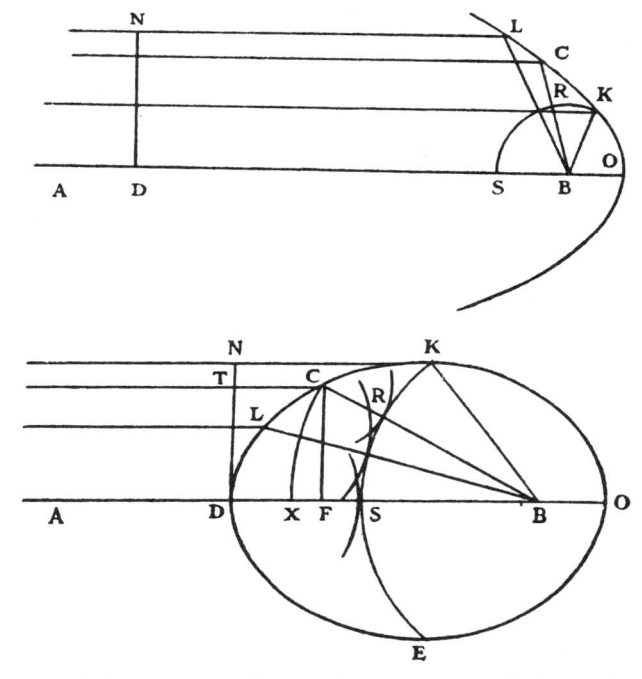

a thing could not arise from the matter of the mirror, and it is entirely impossible.

From what has been demonstrated about this oval, it will be easy to find the figure which serves to collect to a point incident parallel rays. For by supposing just the same construction, but the point A infinitely distant, giving parallel rays, our oval becomes a true Ellipse, the construction

ON LIGHT. Chap. VI

construction of which differs in no way from that of the oval, except that FC, which previously was an arc of a circle, is here a straight line, perpendicular to DB. For the wave of light DN, being likewise represented by a straight line, it will be seen that all the points of this wave, travelling as far as the surface KD along lines parallel to DB, will advance subsequently towards the point B, and will arrive there at the same time. As for the Ellipse which served for reflexion, it is evident that it will here become a parabola, since its focus A may be regarded as infinitely distant from the other, B, which is here the focus of the parabola, towards which all the reflexions of rays parallel to AB tend. And the demonstration of these effects is just the same as the preceding.

But that this curved line CDE which serves for refraction is an Ellipse, and is such that its major diameter is to the distance between its foci as 3 to 2, which is the proportion of the refraction, can be easily found by the calculus of Algebra. For DB, which is given, being called a; its undetermined perpendicular DT being called x; and TC y; FB will be $a-y$; CB will be $\sqrt{xx+aa-2ay+yy}$. But the nature of the curve is such that $\frac{2}{3}$ of TC together with CB is equal to DB, as was stated in the last construction: then the equation will be between $\frac{2}{3}y + \sqrt{xx+aa-2ay+yy}$ and a; which being reduced, gives $\frac{6}{5}ay - yy$ equal to $\frac{9}{5}xx$; that is to say that having made DO equal to $\frac{6}{5}$ of DB, the rectangle DFO is equal to $\frac{9}{5}$ of the square on FC. Whence it is seen that DC is an ellipse, of which the axis DO is to the parameter as 9 to 5; and therefore the square on DO is to the square of the distance between the foci as 9 to $9-5$, that is to say 4; and finally the line DO will be to this distance as 3 to 2.

Again,

Again, if one supposes the point B to be infinitely distant, in lieu of our first oval we shall find that CDE is a true Hyperbola; which will make those rays become parallel which come from the point A. And in consequence also those which are parallel within the transparent body will be collected outside at the point A. Now it must be remarked that CX and KS become straight lines perpendicular to BA, because they represent arcs of circles the centre of which is infinitely distant. And the intersection of the perpendicular CX with the arc FC will give the point C, one of those through which the curve ought to pass. And this operates so that all the parts of the wave of light DN, coming to meet the surface KDE, will advance thence along parallels to KS and will arrive at this straight line at the same time; of which the proof is again the same as that which served for the first oval. Besides one finds by a calculation as easy as the preceding one, that CDE is here a hyperbola of which the axis DO

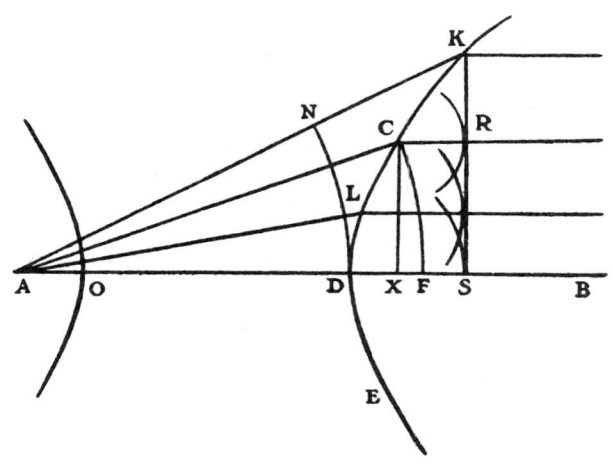

is

ON LIGHT. Chap. VI

is $\frac{4}{5}$ of AD, and the parameter equal to AD. Whence it is easily proved that DO is to the distance between the foci as 3 to 2.

These are the two cases in which Conic sections serve for refraction, and are the same which are explained, in his *Dioptrique*, by Des Cartes, who first found out the use of these lines in relation to refraction, as also that of the

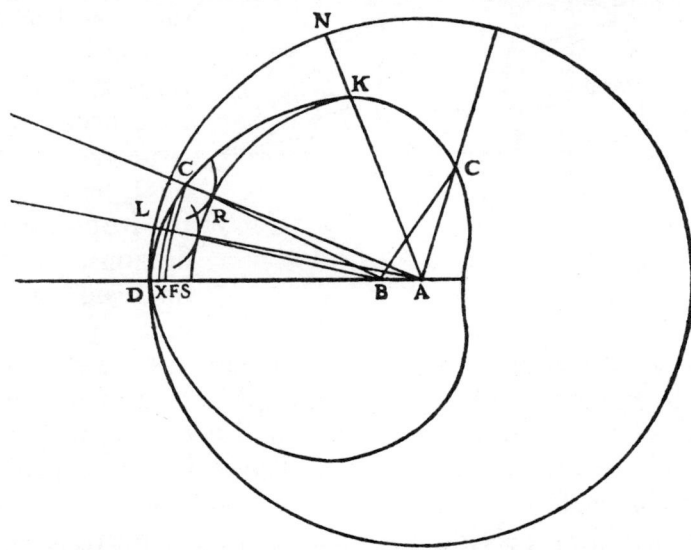

Ovals the first of which we have already set forth. The second oval is that which serves for rays that tend to a given point; in which oval, if the apex of the surface which receives the rays is D, it will happen that the other apex will be situated between B and A, or beyond A, according as the ratio of AD to DB is given of greater or lesser value. And in this latter case it is the same as that which Des Cartes calls his 3rd oval.

Now the finding and construction of this second oval is

the same as that of the first, and the demonstration of its effect likewise. But it is worthy of remark that in one case this oval becomes a perfect circle, namely when the ratio of AD to DB is the same as the ratio of the refractions, here as 3 to 2, as I observed a long time ago. The 4th oval, serving only for impossible reflexions, there is no need to set it forth.

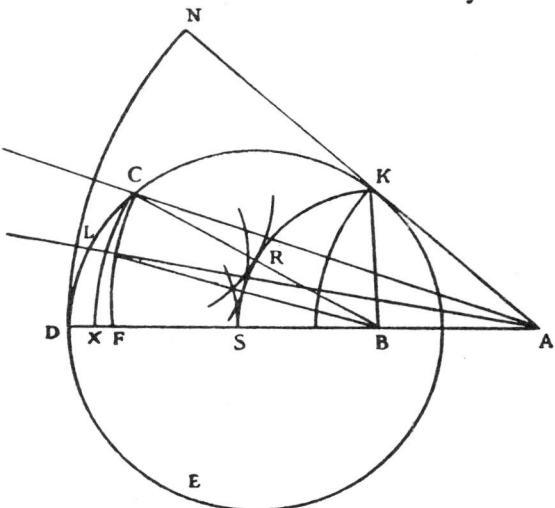

As for the manner in which Mr. Des Cartes discovered these lines, since he has given no explanation of it, nor any one else since that I know of, I will say here, in passing, what it seems to me it must have been. Let it be proposed to find the surface generated by the revolution of the curve KDE, which, receiving the incident rays coming to it from the point A, shall deviate them toward the point B. Then considering this other curve as already known, and that its apex D is in the straight line AB, let us divide it up into an infinitude of small pieces by the points G, C, F; and having drawn from each of these points, straight lines towards A to represent the incident rays, and other straight lines towards B, let there also be described with centre A the arcs GL, CM, FN, DO, cutting the rays that come from A at L, M, N, O; and from the points K, G, C, F,
let

ON LIGHT. Chap. VI 115

let there be described the arcs KQ, GR, CS, FT cutting the rays towards B at Q, R, S, T; and let us suppose that the straight line HKZ cuts the curve at K at right-angles.

Then AK being an incident ray, and KB its refraction within the medium, it needs must be, according to the law

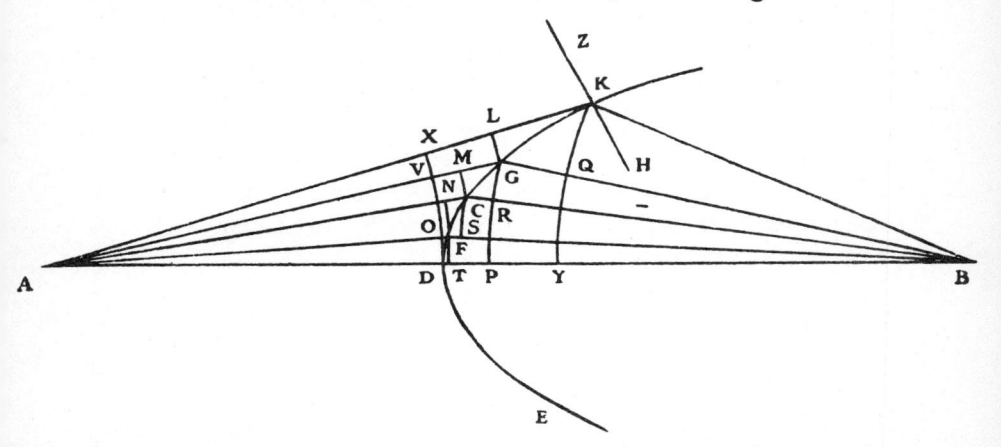

of refraction which was known to Mr. Des Cartes, that the sine of the angle ZKA should be to the sine of the angle HKB as 3 to 2, supposing that this is the proportion of the refraction of glass; or rather, that the sine of the angle KGL should have this same ratio to the sine of the angle GKQ, considering KG, GL, KQ as straight lines because of their smallness. But these sines are the lines KL and GQ, if GK is taken as the radius of the circle. Then LK ought to be to GQ as 3 to 2; and in the same ratio MG to CR, NC to FS, OF to DT. Then also the sum of all the antecedents to all the consequents would be as 3 to 2. Now by prolonging the arc DO until it meets AK at X, KX is the sum of the antecedents. And by prolonging the arc KQ till it meets AD at Y, the sum of the

the consequents is DY. Then KX ought to be to DY as 3 to 2. Whence it would appear that the curve KDE was of such a nature that having drawn from some point which had been assumed, such as K, the straight lines KA, KB, the excess by which AK surpasses AD should be to the excess of DB over KB, as 3 to 2. For it can similarly be demonstrated, by taking any other point in the curve, such as G, that the excess of AG over AD, namely VG, is to the excess of BD over DG, namely DP, in this same ratio of 3 to 2. And following this principle Mr. Des Cartes constructed these curves in his *Geometric*; and he easily recognized that in the case of parallel rays, these curves became Hyperbolas and Ellipses.

Let us now return to our method and let us see how it leads without difficulty to the finding of the curves which one side of the glass requires when the other side is of a given figure; a figure not only plane or spherical, or made by one of the conic sections (which is the restriction with which Des Cartes proposed this problem, leaving the solution to those who should come after him) but generally any figure whatever: that is to say, one made by the revolution of any given curved line to which one must merely know how to draw straight lines as tangents.

Let the given figure be that made by the revolution of some curve such as AK about the axis AV, and that this side of the glass receives rays coming from the point L. Furthermore, let the thickness AB of the middle of the glass be given, and the point F at which one desires the rays to be all perfectly reunited, whatever be the first refraction occurring at the surface AK.

I say that for this the sole requirement is that the outline BDK which constitutes the other surface shall be such

such that the path of the light from the point L to the surface AK, and from thence to the surface BDK, and from thence to the point F, shall be traversed everywhere in equal times, and in each case in a time equal to that which the light employs to pass along the straight line LF of which the part AB is within the glass.

Let LG be a ray falling on the arc AK. Its refraction GV will be given by means of the tangent which will be drawn at the point G. Now in GV the point D must be found such that FD together with $\frac{3}{2}$ of DG and the straight line

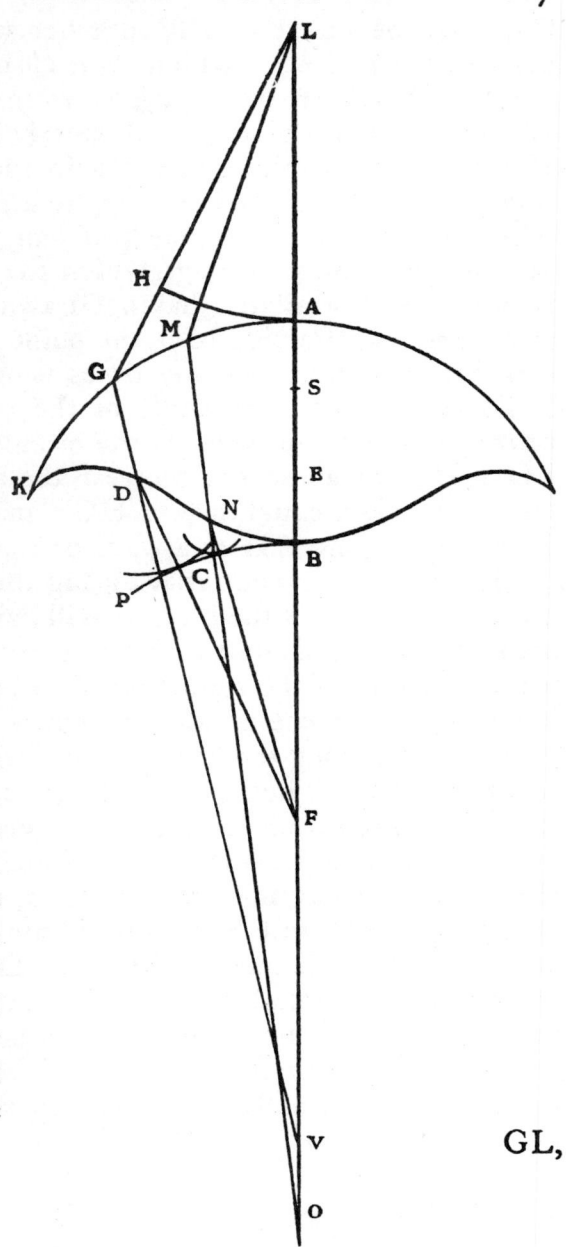

TREATISE

GL, may be equal to FB together with $\frac{3}{2}$ of BA and the straight line AL; which, as is clear, make up a given length. Or rather, by deducting from each the length of LG, which is also given, it will merely be needful to adjust FD up to the straight line VG in such a way that FD together with $\frac{3}{2}$ of DG is equal to a given straight line, which is a quite easy plane problem: and the point D will be one of those through which the curve BDK ought to pass. And similarly, having drawn another ray LM, and found its refraction MO, the point N will be found in this line, and so on as many times as one desires.

To demonstrate the effect of the curve, let there be described about the centre L the circular arc AH, cutting LG at H; and about the centre F the arc BP; and in AB let AS be taken equal to $\frac{2}{3}$ of HG; and SE equal to GD. Then considering AH as a wave of light emanating from the point L, it is certain that during the time in which its piece H arrives at G the piece A will have advanced within the transparent body only along AS; for I suppose, as above, the proportion of the refraction to be as 3 to 2. Now we know that the piece of wave which is incident on G, advances thence along the line GD, since GV is the refraction of the ray LG. Then during the time that this piece of wave has taken from G to D, the other piece which was at S has reached E, since GD, SE are equal. But while the latter will advance from E to B, the piece of wave which was at D will have spread into the air its partial wave, the semi-diameter of which, DC (supposing this wave to cut the line DF at C), will be $\frac{3}{2}$ of EB, since the velocity of light outside the medium is to that inside as 3 to 2. Now it is easy to show that this wave will touch the arc BP at this point C. For since, by construction, FD + $\frac{3}{2}$ DG

$\frac{3}{2}$ DG + GL are equal to FB + $\frac{3}{2}$ BA + AL; on deducting the equals LH, LA, there will remain FD + $\frac{3}{2}$ DG + GH equal to FB + $\frac{3}{2}$ BA. And, again, deducting from one side GH, and from the other side $\frac{3}{2}$ of AS, which are equal, there will remain FD with $\frac{3}{2}$ DG equal to FB with $\frac{3}{2}$ of BS. But $\frac{3}{2}$ of DG are equal to $\frac{3}{2}$ of ES; then FD is equal to FB with $\frac{3}{2}$ of BE. But DC was equal to $\frac{3}{2}$ of EB; then deducting these equal lengths from one side and from the other, there will remain CF equal to FB. And thus it appears that the wave, the semi-diameter of which is DC, touches the arc BP at the moment when the light coming from the point L has arrived at B along the line LB. It can be demonstrated similarly that at this same moment the light that has come along any other ray, such as LM, MN, will have propagated the movement which is terminated at the arc BP. Whence it follows, as has been often said, that the propagation of the wave AH, after it has passed through the thickness of the glass, will be the spherical wave BP, all the pieces of which ought to advance along straight lines, which are the rays of light, to the centre F. Which was to be proved. Similarly these curved lines can be found in all the cases which can be proposed, as will be sufficiently shown by one or two examples which I will add.

Let there be given the surface of the glass AK, made by the revolution about the axis BA of the line AK, which may be straight or curved. Let there be also given in the axis the point L and the thickness BA of the glass; and let it be required to find the other surface KDB, which receiving rays that are parallel to AB will direct them in such wise that after being again refracted at the given surface AK they will all be reassembled at the point L.

From the point L let there be drawn to some point of
the

the given line AK the straight line LG, which, being considered as a ray of light, its refraction GD will then be found. And this line being then prolonged at one side or the other will meet the straight line BL, as here at V. Let there then be erected on AB the perpendicular BC, which will represent a wave of light coming from the infinitely distant point F, since we have supposed the rays to be parallel. Then all the parts of this wave BC must arrive at the same time at the point L; or rather all the parts of a wave emanating from the point L must arrive at the same time at the straight line BC. And for that, it is necessary to find in the line VGD the point D such that having drawn DC parallel to AB, the sum of CD, plus $\frac{3}{2}$ of DG, plus GL may be equal to $\frac{3}{2}$ of AB, plus AL : or rather, on deducting from both sides GL, which is given, CD plus $\frac{3}{2}$ of DG must be equal to a given length; which is a still easier problem than the preceding construction. The point D thus found will be one of those through which the curve ought to pass; and the proof will be the same as before. And by this it will be proved that the waves which come from the point L, after having passed through the glass KAKB, will take

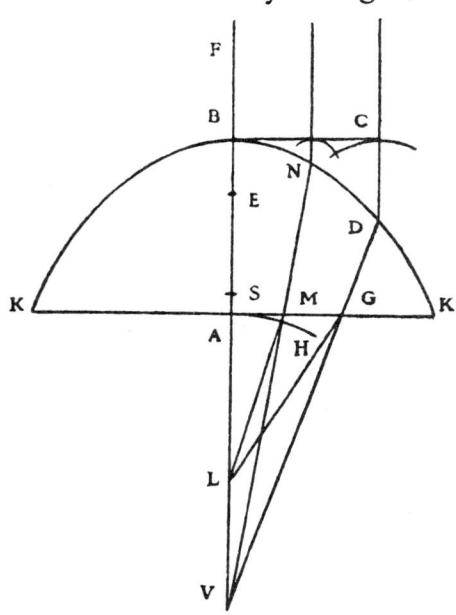

the

the form of straight lines, as BC; which is the same thing as saying that the rays will become parallel. Whence it follows reciprocally that parallel rays falling on the surface KDB will be reassembled at the point L.

Again, let there be given the surface AK, of any desired form, generated by revolution about the axis AB, and let the thickness of the glass at the middle be AB. Also let the point L be given in the axis behind the glass; and let it be supposed that the rays which fall on the surface AK tend to this point, and that it is required to find the surface BD, which on their emergence from the glass turns them as if they came from the point F in front of the glass.

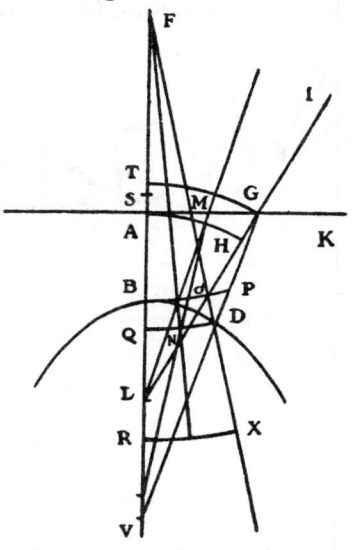

Having taken any point G in the line AK, and drawing the straight line IGL, its part GI will represent one of the incident rays, the refraction of which, GV, will then be found: and it is in this line that we must find the point D, one of those through which the curve DG ought to pass. Let us suppose that it has been found: and about L as centre let there be described GT, the arc of a circle cutting the straight line AB at T, in case the distance LG is greater than LA; for otherwise the arc AH must be described about the same centre, cutting the straight line LG at H. This arc GT (or AH, in the other case) will represent an incident wave of light, the rays of which
tend

tend towards L. Similarly, about the centre F let there be described the circular arc DQ, which will represent a wave emanating from the point F.

Then the wave TG, after having passed through the glass, must form the wave QD; and for this I observe that the time taken by the light along GD in the glass must be equal to that taken along the three, TA, AB, and BQ, of which AB alone is within the glass. Or rather, having taken AS equal to $\frac{2}{3}$ of AT, I observe that $\frac{3}{2}$ of GD ought to be equal to $\frac{3}{2}$ of SB, plus BQ; and, deducting both of them from FD or FQ, that FD less $\frac{3}{2}$ of GD ought to be equal to FB less $\frac{3}{2}$ of SB. And this last difference is a given length: and all that is required is to draw the straight line FD from the given point F to meet VG so that it may be thus. Which is a problem quite similar to that which served for the first of these constructions, where FD plus $\frac{3}{2}$ of GD had to be equal to a given length.

In the demonstration it is to be observed that, since the arc BC falls within the glass, there must be conceived an arc RX, concentric with it and on the other side of QD. Then after it shall have been shown that the piece G of the wave GT arrives at D at the same time that the piece T arrives at Q, which is easily deduced from the construction, it will be evident as a consequence that the partial wave generated at the point D will touch the arc RX at the moment when the piece Q shall have come to R, and that thus this arc will at the same moment be the termination of the movement that comes from the wave TG; whence all the rest may be concluded.

Having shown the method of finding these curved lines which serve for the perfect concurrence of the rays,
there

there remains to be explained a notable thing touching the uncoordinated refraction of spherical, plane, and other surfaces : an effect which if ignored might cause some doubt concerning what we have several times said, that rays of light are straight lines which intersect at right angles the waves which travel along them.

For in the case of rays which, for example, fall parallel upon a spherical surface AFE, intersecting one another, after refraction, at different points, as this figure represents; what can the waves of light be, in this transparent body, which are cut at right angles by the converging rays? For they can not be spherical. And what will these waves become after the said rays begin to intersect one another? It will be seen in the solution of this difficulty that something very remarkable comes to pass herein, and that the waves do not cease to persist though they do not continue entire, as when they cross the glasses designed according to the construction we have seen.

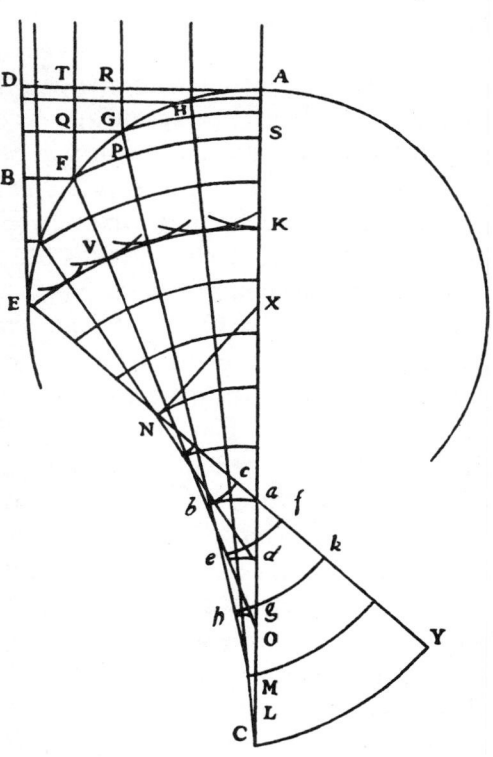

According

According to what has been shown above, the straight line AD, which has been drawn at the summit of the sphere, at right angles to the axis parallel to which the rays come, represents the wave of light; and in the time taken by its piece D to reach the spherical surface AGE at E, its other parts will have met the same surface at F, G, H, etc., and will have also formed spherical partial waves of which these points are the centres. And the surface EK which all those waves will touch, will be the continuation of the wave AD in the sphere at the moment when the piece D has reached E. Now the line EK is not an arc of a circle, but is a curved line formed as the evolute of another curve ENC, which touches all the rays HL, GM, FO, etc., that are the refractions of the parallel rays, if we imagine laid over the convexity ENC a thread which in unwinding describes at its end E the said curve EK. For, supposing that this curve has been thus described, we will show that the said waves formed from the centres F, G, H, etc., will all touch it.

It is certain that the curve EK and all the others described by the evolution of the curve ENC, with different lengths of thread, will cut all the rays HL, GM, FO, etc., at right angles, and in such wise that the parts of them intercepted between two such curves will all be equal; for this follows from what has been demonstrated in our treatise *de Motu Pendulorum*. Now imagining the incident rays as being infinitely near to one another, if we consider two of them, as RG, TF, and draw GQ perpendicular to RG, and if we suppose the curve FS which intersects GM at P to have been described by evolution from the curve NC, beginning at F, as far as which the thread is supposed to extend, we may assume the small piece FP as a straight line perpendicular

to

to the ray GM, and similarly the arc GF as a straight line. But GM being the refraction of the ray RG, and FP being perpendicular to it, QF must be to GP as 3 to 2, that is to say in the proportion·of the refraction; as was shown above in explaining the discovery of Des Cartes. And the same thing occurs in all the small arcs GH, HA, etc., namely that in the quadrilaterals which enclose them the side parallel to the axis is to the opposite side as 3 to 2. Then also as 3 to 2 will the sum of the one set be to th sum of the other; that is to say, TF to AS, and DE to AK, and BE to SK or DV, supposing V to be the intersection of the curve EK and the ray FO. But, making FB perpendicular to DE, the ratio of 3 to 2 is also that of BE to the semi-diameter of the spherical wave which emanated from the point F while the light outside the transparent body traversed the space BE. Then it appears that this wave will intersect the ray FM at the same point V where it is intersected at right angles by the curve EK, and consequently that the wave will touch this curve. In the same way it can be proved that the same will apply to all the other waves above mentioned, originating at the points G, H, etc.; to wit, that they will touch the curve EK at the moment when the piece D of the wave ED shall have reached E.

Now to say what these waves become after the rays have begun to cross one another: it is that from thence they fold back and are composed of two contiguous parts, one being a curve formed as evolute of the curve ENC in one sense, and the other as evolute of the same curve in the opposite sense. Thus the wave KE, while advancing toward the meeting place becomes *abc*, whereof the part *ab* is made by the evolute *b*C, a portion of the curve ENC,

ENC, while the end C remains attached; and the part *bc* by the evolute of the portion *b*E while the end E remains attached. Consequently the same wave becomes *def*, then *ghk*, and finally CY, from whence it subsequently spreads without any fold, but always along curved lines which are evolutes of the curve ENC, increased by some straight line at the end C.

There is even, in this curve, a part EN which is straight, N being the point where the perpendicular from the centre X of the sphere falls upon the refraction of the ray DE, which I now suppose to touch the sphere. The folding of the waves of light begins from the point N up to the end of the curve C, which point is formed by taking AC to CX in the proportion of the refraction, as here 3 to 2.

As many other points as may be desired in the curve NC are found by a Theorem which Mr. Barrow has demonstrated in section 12 of his *Lectiones Opticae*, though for another purpose. And it is to be noted that a straight line equal in length to this curve can be given. For since it together with the line NE is equal to the line CK, which is known, since DE is to AK in the proportion of the refraction, it appears that by deducting EN from CK the remainder will be equal to the curve NC.

Similarly the waves that are folded back in reflexion by a concave spherical mirror can be found. Let ABC be the section, through the axis, of a hollow hemisphere, the centre of which is D, its axis being DB, parallel to which I suppose the rays of light to come. All the reflexions of those rays which fall upon the quarter-circle AB will touch a curved line AFE, of which line the end E is at the focus of the hemisphere, that is to say, at the point which divides the semi-diameter BD into two equal parts. The

The points through which this curve ought to pass are found by taking, beyond A, some arc AO, and making the arc OP double the length of it; then dividing the chord OP at F in such wise that the part FP is three times the part FO; for then F is one of the required points.

And as the parallel rays are merely perpendiculars to the waves which fall on the concave surface, which waves are parallel to AD, it will be found that as they come successively to encounter the surface AB, they form on reflexion folded waves composed of two curves which originate from two opposite evolutions of the parts of the curve AFE. So, taking AD as an incident wave, when the part AG shall have met the surface AI, that is to say when the piece G shall have reached I, it will be the curves HF, FI, generated as evolutes of the curves FA, FE, both beginning at F, which together constitute the propagation of the part AG. And a little afterwards, when the part AK has met the surface AM, the piece K having come to M, then the curves LN, NM, will together constitute the propagation of that part. And thus this folded wave will continue to advance until the point N has reached the focus E. The curve AFE can be seen in smoke, or in flying dust, when a concave mirror is held opposite the sun. And it should be known that it is none other than that curve which is described

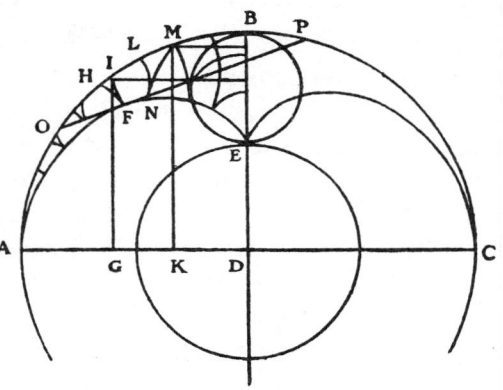

scribed by the point E on the circumference of the circle EB, when that circle is made to roll within another whose semi-diameter is ED and whose centre is D. So that it is a kind of Cycloid, of which, however, the points can be found geometrically.

Its length is exactly equal to $\frac{3}{4}$ of the diameter of the sphere, as can be found and demonstrated by means of these waves, nearly in the same way as the mensuration of the preceding curve; though it may also be demonstrated in other ways, which I omit as outside the subject. The area AOBEFA, comprised between the arc of the quarter-circle, the straight line BE, and the curve EFA, is equal to the fourth part of the quadrant DAB.

END.

INDEX

Archimedes, 104.
Atmospheric refraction, 45.
Barrow, Isaac, 126.
Bartholinus, Erasmus, 53, 54, 57, 60, 97, 99.
Boyle, Hon. Robert, 11.
Cassini, Jacques, iii.
Caustic Curves, 123.
Crystals, see Iceland Crystal, Rock Crystal.
Crystals, configuration of, 95.
Descartes, René, 3, 5, 7, 14, 22, 42, 43, 109, 113.
Double Refraction, discovery of, 54, 81, 93.
Elasticity, 12, 14.
Ether, the, or Ethereal matter, 11, 14, 16, 28.
Extraordinary refraction, 55, 56.
Fermat, principle of, 42.
Figures of transparent bodies, 105.
Hooke, Robert, 20.
Iceland Crystal, 2, 52 sqq.
Iceland Crystal, Cutting and Polishing of, 91, 92, 98.
Leibnitz, G. W., vi.
Light, nature of, 3.

Light, velocity of, 4, 15.
Molecular texture of bodies, 27, 95.
Newton, Sir Isaac, vi, 106.
Opacity, 34.
Ovals, Cartesian, 107, 113.
Pardies, Rev. Father, 20.
Rays, definition of, 38, 49.
Reflexion, 22.
Refraction, 28, 34.
Rock Crystal, 54, 57, 62, 95.
Römer, Olaf, iii, 7.
Roughness of surfaces, 27.
Sines, law of, 1, 35, 38, 43.
Spheres, elasticity of, 15.
Spheroidal waves in crystals, 63.
Spheroids, lemma about, 103.
Sound, speed of, 7, 10, 12.
Telescopes, lenses for, 62, 105.
Torricelli's experiment, 12, 30.
Transparency, explanation of, 28, 31, 32.
Waves, no regular succession of, 17.
Waves, principle of wave envelopes, 19, 24.
Waves, principle of elementary wave fronts, 19.
Waves, propagation of light as, 16, 63.